THE

# NATURAL HISTORY

OF

# BRITISH INSECTS;

EXPLAINING THEM

IN THEIR SEVERAL STATES,

WITH THE PERIODS OF THEIR TRANSFORMATIONS,
THEIR FOOD, OECONOMY, &c.

TOGETHER WITH THE

## HISTORY OF SUCH MINUTE INSECTS

AS REQUIRE INVESTIGATION BY THE MICROSCOPE.

THE WHOLE ILLUSTRATED BY

# COLOURED FIGURES,

DESIGNED AND EXECUTED FROM LIVING SPECIMENS.

By E. DONOVAN.

VOL. VI.

LONDON:

PRINTED FOR THE AUTHOR,

And for F. and C. RIVINGTON, N° 62, St. PAUL's CHURCH-YARD.

MDCCXCVII.

THE

# NATURAL HISTORY

OF

# BRITISH INSECTS.

---

## PLATE CLXXXI.

### PHALÆNA SYRINGARIA.

#### LEPIDOPTERA.

#### *GENERIC CHARACTER.*

Antennæ taper from the bafe. Wings in general deflexed when at reft. Fly by night.

#### GEOMETRA.

#### *SPECIFIC CHARACTER*

AND

#### *SYNONYMS.*

Wings angulated, indented, fine light grey, with fhades of flefh colour and brown: a brown mark acrofs the Wings, which in an expanded Infect forms a feftoon.

PHALÆNA SYRINGARIA: pectinicornis, alis fuberofis, omnibus grifeo flavefcentibus, ftrigis repandis fufcis albifque. *Fab. Syft. Ent.* 622. 13.—*Spec. Inf.* 2. 244. 17.— *Linn. Syft. Nat.* 2. 860. 206.

B 2

*Phalæna*

*Phalæna* pectinicornis, alis margine finuatus, fulvo, nigro, fufco
　　　rofeoque marmoratis, fingulis fubtus puncto nigro,
　　　fuperioribus extremo dilatato, recurvis. *Geoff. Inf.*
　　　2. 126. 32.
　　　*Roef. Inf.* 1. *phal.* 3. *tab.* 10.

———————————————

The Englifh name has been given to this Moth by fome collectors,
on account of its being found a local fpecies, and moft frequent in
Richmond park.

The Larva of this Infect has a very fingular appearance, to which
the hook on the back greatly contributes when it is nearly full fed
and going into the pupa ftate. It feeds on the Jafmine and Lilac,
and does not thrive well on any other plants. It is in the Caterpillar
ftate in April, changes to chryfalis in May: the Fly comes forth in
June.

PLATE

1

2

1

# PLATE CLXXXII.

## MELOE MONOCEROS.

### HORNED MELOE.

### COLEOPTERA.

### *GENERIC CHARACTER.*

Antennæ globular, laft globule oval. Thorax roundifh. Elytra foft. Head gibbous.

### *SPECIFIC CHARACTER*

#### AND

### *SYNONYMS.*

Thorax pointed like a horn over the head. Shells brown, with a longitudinal ftreak and fpot on each.

MELOE MONOCEROS : *Linn. Syft. Nat.* 2. 681. 14.
NOTOXUS MONOCEROS : thorace in cornu fupra caput protenfo, elytris puncto fafciaque nigris, *Fab. Syft. Ent. Tom.* 1. 211. 6.
Notoxus. *Geoff. Inf.* 1. 356. *tab.* 6. *fig.* 8.
Attelabus Monoceros. *Lin. Fn. Sv.* 638.
*Schæff. Icon. tab.* 188. *fig.* 3.

---

Linnæus feems to have found much difficulty in determining the genus of this infect; once placing it amongft the Attelabi of his fyftem, and thence removing it to the Meloe tribe. Geoffroy has another generical name, *notoxus*, and this Fabricius has adopted in his laft work.

This

This is certainly a moſt ſingular Inſect, and altogether unlike any other we know of in England. We cannot ſay whether it feeds on any particular food, or what are its peculiarities, for we have only ſeen one ſpecimen beſides that from which the annexed drawing is taken; and the only information we receive from the authors above quoted is, that it is a native of Europe; and is found on umbelliferous plants. It was taken in May.

Fig. 1. the natural ſize. Fig. 2. magnified.

PLATE

# 183

# PLATE CLXXXIII.

## PHALÆNA CAMELINA.

### DARK PROMINENT MOTH.

### LEPIDOPTERA.

### *GENERIC CHARACTER.*

Antennæ taper from the bafe. Wings in general deflexed when at reft. Fly by night.

*Bombyx.*

### *SPECIFIC CHARACTER*

AND

### *SYNONYMS.*

Firft wings fine ruft colour, with two oblique waves acrofs: indented edges: pofterior margin, with one large and one fmall tuft prominent on the upper part of the Infect when at reft.

PHALÆNA CAMELINA: alis deflexis denticulatis brunneis: omnibus denticulo dorfali. *Linn. Syft. Nat.* 2. 832. 80.—*Fn. Sv.* 1145.
*Fab. Ent. Syft.* 3. *p.* 450. 133.
*Wien. Verz.* 63. 3.
*Roef. Inf.* 1. *phal.* 2. *tab.* 28.

---

The Caterpillar of this Moth is found in Auguft, on the leaves of the Oak, Willow, Lime, &c. The Moth comes forth late in May, or early in June.

C　　　　　　　　　　PLATE

# 184

# PLATE CLXXXIV.

## PAPILIO ARION.

### MAZARINE BLUE BUTTERFLY.

### LEPIDOPTERA.

### *GENERIC CHARACTER.*

Antennæ clubbed. Wings when at rest, erect. Fly by day.

### *SPECIFIC CHARACTER*

### AND

### *S Y N O N Y M S.*

Above brown, disk of the wings fine blue, with black spots. Beneath grey, with a number of small eyes.

PAPILIO ARION: alis supra fuscus: disco cœruleo; maculis atris,
        subtus canis: punctis ocellaribus. *Linn. Syst. Nat.* 2.
        789. 230.—*Fn. Sv.* 1073.
Hesperia Arion. *Fab. Ent. Syst.* 3. 293. 118.
        *Reef. Inf.* 3. *tab.* 45. *fig.* 3. 4.
        *Schæff. Icon. tab.* 98. *fig.* 5. 6.
        *Esp. pap.* 1. *tab.* 20. *fig.* 2.

---

Papilio Arion is a very scarce Insect in this country; and it does not appear to be much more common in any other part of Europe. The authors quoted above have given figures or descriptions of it, but are entirely ignorant of its larva, or pupa; and Fabricius, who has selected the observations of all the preceding authors who have described it, says only, *Habitat in Europæ Pratis.*

C 2

Mr.

Mr. Lemon, a collector of eminence fome years fince, met with it in England. We have a fpecimen which was taken by him, as appears from a note with it. Roefel's figure is of a finer blue than any we have feen ; but we have no doubt of its being a moft brilliant Infect in a recent ftate.

PLATE

# 185

1

1

2

2

# PLATE CLXXXV.

## FIG. I.

### CHRYSOMELA CALMARIENSIS.

#### COLEOPTERA.

*GENERIC CHARACTER.*

Antennæ knotted, enlarging towards the ends. Corfelet margined.

*SPECIFIC CHARACTER*

AND

*SYNONYMS.*

Oblong: yellow: a longitudinal ftreak of black down each fhell.

CHRYSOMELA CALMARIENSIS: oblongiufcula ferruginea, elytris
macula longitudinali nigricante. *Linn. Syft. Nat.* 2.
600. 101.
*Crioceris Calmarienfis*: ovata cinerea, elytris vitta lineolaque bafeos
nigris. *Fab. Syft. Ent.* 119. 4.—*Spec. Inf.* 1. 150. 6.
*Galleruca* pallida, thorace nigro variegato, elytris fafciis duabus longitudinalibus nigris. *Geoff. Inf.* 1. 253. 3.—*Sulz.
Roem. Inf. tab.* 3. *fig.* 16.

---

Fabricius fays, this Infect lives on the Willow and Alder. It is
found in every part of Europe, but is fcarce in England.

It is a very pretty fpecies. The underfide has rather a bronze
appearance: above, in recent fpecimens, the light colour is very

C 3                                                                                      fine

fine yellow, but turns brown afterwards: the ſtripes down the elytra are not deep black, but have a greeniſh tint.—Taken in May. Length a quarter of an inch.

---

## F I G. II.

## MYCETOPHAGUS QUADRIMACULATUS.

Fungus Beetle with four Spots.

Coleoptera.

### GENERIC CHARACTER.

*Palpi* four, unequal. *Maxilla* membranaceous, with a tooth or ſpine. Lip round, entire. Antennæ gradually enlarge towards the end. *Fab. Gen. Inſ.*

### SPECIFIC CHARACTER

Entirely black, except one large yellow brown ſpot on the baſe of the elytra, and another near their extremities.

Mycetophagus Quadrimaculatus: rufus thorace elytriſque
   nigris, his maculis duabus rufis. *Fab. Ent. Syſt. t.* **2.**
   *n.* 1. *p.* 497.
Ips Maculata. *Fab. Mant. Inſ.* 1. *n.* 8. *p.* 45.
Tritoma. *Geoffr. Inſ. t.* 1. *p.* 335. *tab.* 6. *fig.* 2. *e. f.*
Silphoides boleti. *Herbſt. Archiv.* 4 *tes. Heft.* 2. 11. *p.* 41. *tab.* 21.
   *fig.* 51.
Der Viergefleckte Pſiſſerkäfer. *Panz. Faun. Inſ. Germ.*

---

Three ſpecimens of this rare ſpecies were found together, concealed in the porous part of the Honey-comb Boletus *. According to *Herbſt. Archiv.* it is uſually found on plants of this genus. It is not ſcarce in Germany.

---

\* Boletus Celluloſus.

The

PLATE CLXXXV. 15

The vaſt addition of new ſpecies of Inſects that *Fabricius* has made in his latter works, renders it difficult, and in ſome inſtances impoſſible, to refer them to the ſyſtem of *Linnæus* ; and though we would at all times more willingly refer any new kind to that ſyſtem, and quote the Fabrician account amongſt the ſynonyms, than adopt the new generic diſtinctions of Fabricius, or any other author, we muſt not ſacrifice propriety to predilection. If Linnæus himſelf had noticed many of the newly diſcovered kinds, he muſt have altered and improved his ſyſtem to admit them ; which is a great excuſe for Fabricius having made ſo many genera.

The little Inſect before us may be referred to the Silpha genus of Linnæus ; but as his definition of the Silphæ has been ſubjected to ſo many objections and amendments of later authors, we forbear placing it amongſt them. Linnæus has made no diſtinction between ſuch ſpecies as have perfoliated antennæ, and thoſe with antennæ increaſing in bulk from the baſe to the end †, thereby including *S. Veſpillo* with *S. thoracica* and *S. obſcura,* though in an early edition of the *Fauna Suecica,* *S. Veſpillo* was amongſt the *Scarabæi,* and the two laſt with the *Caſſidæ.*

*Schœffer* divided the Silphæ of Linnæus into two genera, *Silpha* and *Peltis.* *Geoffroy* arranged ſome of them amongſt his *Dermeſtides,* and formed his genus *Peltis* of ſuch as have the thorax and elytra ſtrongly margined, and perfoliated antennæ. *Geoffroy* deſcribed our preſent ſubject under the name *Tritoma.* *Fabricius* afterwards, in his Mantiſſa, arranged it under *Ips,* a new genus, formed from part of the *Silphæ* of Linnæus and *Dermeſtides* of Geoffroy. In the Entomologia Syſtematica, he has again ſeparated the Ips, and made *Mycetophagus* a new genus of fifteen ſpecies, in which he includes this Inſect.

---

† *Silpha Antennæ* extrorſum craſſiores. *Elytra* marginata. *Caput* prominens. *Thorax* planiuſculus, marginatus. *Linn. Gen. Inſ.*

To thofe who have adopted the fyftem of Fabricius, it is unne-
ceffary to fay, that the fundamental part of his arrangement is taken
from the mouth of the Infect, which certainly is objectionable, becaufe that part is dry, and very frequently injured or deftroyed in
Infects that have been long in a cabinet, and is very fmall in moft
kinds when alive ; fo that, though we have followed his arrangement
in the generic definition, we have been lefs prolix than a full elucidation of it might require.

The figures in the annexed plate 2. 2. reprefent the natural fize
and magnified appearance.

PLATE

# PLATE CLXXXVI.

## PAPILIO HERO.

### SCARCE MEADOW-BROWN BUTTERFLY.

#### LEPIDOPTERA.

### GENERIC CHARACTER.

Antennæ clubbed. Wings, when at reſt, erect. Fly by day.

### SPECIFIC CHARACTER

#### AND

### SYNONYMS.

Wings entire : brown above. Beneath, a large black eye near the anterior margin of the firſt wings : ſix ſmaller eyes on the ſecond Wings.

PAPILIO HERO: alis integerrimis fulvis: ſubtus anticis ocello, porticis ſenis. *Linn. Syſt. Nat.* 2. 793. 253.—*Fn. Sv.* 1047.—*Fab. Ent. Syſt.* 3. *p.* 222. 695.
Papilio Hero. *Wien. Verz.* 168. 13.
Papilio Typhon. *Eſp. pap. tab.* 35. *fig.* 3. 4.

---

This is a local ſpecies : it is very abundant in ſome marſhy parts of Lancaſhire ; but we have not learnt that it has bęen taken in any other part of the kingdom. Many of the curious in London are particularly indebted to Mr. Phillips, of Mancheſter, for enriching their cabinets with *Papilio Hero* ; for, though it is a plain Inſect, it is eſteemed for its rarity, few Entomologiſts having travelled into that part of the country to collect Inſects.

PLATE

# PLATE CLXXXVII.

## PHALÆNA EXOLETA.

### Sword Grass Moth.

### Lepidoptera.

## GENERIC CHARACTER.

Antennæ taper from the bafe. Wings in general deflexed when at reft. Fly by night.

*Noctua.*

Antennæ in both fexes, like a briftle.

## SPECIFIC CHARACTER

### AND

### SYNONYMS.

Thorax crefted. Wings lance-fhaped, varied with grey and brown, a kidney-fhaped fpot in the middle: four white fpots on the anterior margin.

PHALÆNA EXOLETA : criftata, alis lanceolatis convolutis fufco cinereoque nebulofis, punctis quatuor marginalibus albis. *Fab. Syft. Ent.* 617. 116.—*Spec. Inf.* 2. 239. 144.

---

The beautiful Caterpillar of this Moth is found on Sword Grafs in Auguft. Though its trivial name implies that it is peculiar to

this

this plant, feveral others are mentioned by authors as proper food for it, amongft thefe are the Bell flower \*, Orach †, and common Pea. The Moth appears in May: frequents marfhy places.

---

\* Campanula.       † Atriplex.

**PLATE**

# 188

# PLATE CLXXXVIII.

## HEMEROBIUS CHRYSOPS.

### NEUROPTERA.

Wings four, tranfparent, reticulated. Tail without a fting.

### *GENERIC CHARACTER.*

Mouth armed with two teeth and four palpi. Wings deflected. Antennæ fetaceous. Thorax convex.

### *GENERIC CHARACTER*

### AND

### *SYNONYMS.*

Greenifh. Wings tranfparent, reticulated, with dark fpots.

HEMEROBIUS CHRYSOPS: viridi nigroque varius, alis hyalinis: venis viridibus; lineolis nigris reticulatis. *Linn. Syft. Nat.* 2. 912. 4.—*Fn. Sv.* 1505.
*Geoff. Inf.* 2. 254. 2.
*Fab. Ent. Syft. t.* 2. 83. *f.* 6.
*Frifch. Inf.* 4. 40. *tab.* 23.
*Roef. Inf.* 3. *tab.* 21. *fig.* 4.
*Sulz. Hift. Inf. tab.* 25. *fig.* 1.

This Infect was formerly held in great efteem amongft the Englifh collectors, on account of its rarity; and has been purchafed for their cabinets at a confiderable price *. The late Mr. Bentley, who

---

* Half a guinea was the ufual price for a pair of them.

had

had been more than twenty years endeavouring to make his cabinet the moſt complete in England, never met with it.

About three years ſince, they were taken in great plenty near London, both at Batterſea and Highgate ; and have been met with in other places ſince that time. Like the Ephemeræ, and other gauſe-wing Inſects, it delights in moiſt places, particularly among the reeds. The larva is unknown, but we ſuppoſe that it lives in that ſtate in the water ; and which moſt likely it does not leave till it becomes a winged creature. The larva of ſome ſpecies of this genus feed on the ſmaller kinds of Inſects.

The wings are the moſt pleaſing objects for the microſcope that can be imagined: the reticulations and feathered edges are ſo tranſ parent, that they may be examined with the deepeſt magnifiers, which is an advantage few objects of ſuch a ſize poſſeſs. The magnified figure is given, with the natural ſize in the annexed plate.

Fabricius has erroneouſly quoted the ſeventh and eighth figure of the fifth plate of *Schæffer's Icones*, which is certainly no other than the common kind, Hemorobius perla, and which is ſo often found in gardens with a fine golden eye. The figure quoted in *Sulzer* and *Roeſel* agree with our ſpecimen.

**P L A T E**

# PLATE CLXXXIX.

PHALÆNA MENTHRASTIRI.

SPOTTED WHITE MOTH.

LEPIDOPTERA.

## GENERIC CHARACTER.

Antennæ taper from the bafe. Wings in general deflexed when at reft. Fly by night.

*Bombyx.*

## SPECIFIC CHARACTER

AND

## SYNONYMS.

White with black fpots. Abdomen orange, with black fpots.

PHALÆNA MENTHRASTRI: alis deflexis albis nigro fubpunctatis, abdominis dorfo fulvo nigro punctato, femoribus anticis luteis. *Fab. Ent. Syft. T.* 3. *p.* 1. 452. 140.
Bombyx Menthraftri. *Wien. Verz* 54. 2.—*Roef. Inf.* 1 *Phal.* 2. *t.* 46.
*Knoch. Beytr.* 3. *tab.* 2. *fig.* 5. 13.

---

This Infect has been confounded with Phalæna lubricipeda by Linnæus; he makes it the variety β after De Geer. In this he has been followed by many other authors; and though Roefel, by giving the larva and pupa of each, in two diftinct plates, evidently thought them different fpecies, his obfervations had no weight with other Naturalifts; even Fabricius, in his *Species Infectorum*, gives them under

under one fpecific name.  In his laft work, *Entomologia Syftematica*, he has divided them, leaving the P. lubricipeda under its former name, and giving the fpecific name Menthraftri to the white fort, as had been done in *Wien. Verz.* 54. 2.  Fabricius mentions it as a native of Germany, but from the figure of Roefel no doubt can be entertained of its being precifely the fame as our Englifh fpecies.

The Caterpillars of both forts are very general feeders ; they will eat, oak, fruit trees, and wild plants of almoft every kind.  They are common in the fummer, change to chryfalis about Auguft, and appear in the winged ftate in May and June ; but, there is more than one brood of them in the courfe of the year, fo that the time of their appearance is uncertain.  The Caterpillars change their fkins often ; and change their colours at the fame time.  Thofe of Phalæna Menthraftri when fmall are a very light tranfparent brown : then brown with dark ftripes.  It is not black till it is in the laft fkin ; and then, in many, the colour inclines to brown.

PLATE

# PLATE CXC. CXCI.

## SPHINX CELERIO.

### SILVER-STRIPE HAWK MOTH.

#### LEPIDOPTERA.

## GENERIC CHARACTER.

Antennæ thickeſt in the middle. Wings, in general deflexed when at reſt. Fly ſlow. Morning and evening only.

## SPECIFIC CHARACTER

### AND

## SYNONYMS.

Firſt wings brown, with a broad oblique band of ſilver white extending from the poſterior margin to the tip of the wing. Lower wings black, with ſix large red ſpots on each.

Sphinx Celerio: alis integris griſeis: ſtriis albis, poſticis fuſcus: maculis ſex rubris. *Linn. Syſt. Nat.* 2. 800. 12.
   *Fab. Ent. Syſt. T. p.* 1. 370. 43.
   *Roeſ. Inſ.* 3. *tab.* 8.
   *Friſch. Inſ.* 13. *tab.* 1. *fig.* 2.
   *Cram. Inſ.* 3. *tab.* 25. *B.*

---

The Sphinx Celerio ſtands pre-eminent in the liſt of the Inſects of this country, whether we conſider its rarity, or uncommon beauty. Indeed, amongſt the Inſects of this tribe that are brought from remote countries, even from Aſia, which boaſts the moſt ſplendid ſpecies, the varieties of Sphinx Celerio are often the moſt beautiful ; it muſt however be owned, that, in countries where the

D          climate,

climate, and luxuriance of the foil contribute to enrich the juices of the plants on which the Infects are nourifhed, they are larger, and their colours more vivid than any of the fame kind produced in the northern countries of Europe.

Several years fince, Mr. Francillon, of Norfolk-ftreet in the Strand, had a living fpecimen of this Infect brought to him : it was taken in Bunhill-fields burying-ground. It is ftill preferved in his cabinet.

We have heard of other fpecimens being taken in this country ; but the only inftance we can quote with confidence is, that Mr. Latham, formerly of Dartford, and well known for his fcientific refearches in natural hiftory, has a fpecimen which was taken at Eltham, in Kent. Few collections of confequence are without this Infect, but they are in general natives of Germany.

Roefel has given a figure of this Sphinx, with its larva and pupa ; and, as we could never reafonably expect to meet with it in thefe ftates in England, correct copies of his figures are given in Plate 191. The works of Roefel are not in the hands of many ; and, we are convinced, that Plate will be acceptable to moft of our readers.

Sphinx Celerio is found on the vine.

PLATE

# PLATE CXCII.

## PHALÆNA HEXAPTERATA.

SERAPHIM MOTH.

LEPIDOPTERA.

### GENERIC CHARACTER.

Antennæ taper from the bafe. Wings in general deflexed when at reft. Fly by night.

### SPECIFIC CHARACTER

AND

### SYNONYMS.

Firft wings varied with brown and grey. Second pair white, with an appendage at the bafe of each, refembling a fmall wing.

PHALÆNA HEXAPTERA: feticornis alis fufco grifeoque variis: pofticis albis bafi duplicatis. *Fab. Ent. Syft. Tom.* 3. *p.* 2. 193. 233.

Phalæna Hexapterata. *Wien. Verz.* 109. 10.

─────────

Fabricius has given an accurate defcription* of this extraordinary Infect; but he makes no reference to any work that contains a figure of it. The nineteenth plate of Kleman's Continuation of Roeffel's *Infecten-Beluftigungen*, entitled *Infecten-Gefchichte*, certainly

─────────

* Media. Alæ anticæ grifeæ, fufco undatae imprimis bafi apiceque, in medio parum dilutiores lunula fufca. Pofticæ albae ad bafin ala notha, rotundatà, alba, immaculata. Subtus omnes cinereæ puncto medio, fufco.—*Fab.*

E          efcaped

efcaped his notice; for in that plate we find both fexes, Figures *a*,
*b*: from this account we learn that Phalæna Hexapterata is a native
of Germany. The female has four wings: the male appears at
firft fight to have fix, which is more than any tribe of Infects are
furnifhed with; a fmall appendage very much refembling a wing,
and of the fame texture, arifes from the bafe of the fecond pair of
wings next the abdomen. The nerves of the true wing ramify into
this appendage; which when the Infect is expanded, give it moft
fingular appearance. In the annexed plate this appendage is magnified, to enable us more accurately to exhibit its true form and
tendons.

Our fpecimen was taken on Epping Foreft in 1795. It is very
rare in England. Fabricius fays it feeds on the Beech*.

---

* Habitat in Fago Sylvatica.

PLATE

# 193

# PLATE CXCIII.

## PHALÆNA PYRAMIDEA.

### COPPER-UNDERWING MOTH.

### LEPIDOPTERA.

### *GENERIC CHARACTER.*

Antennæ taper from the bafe. Wings in general deflexed when at reft. Fly by night.

### *SPECIFIC CHARACTER*

### AND

### *SYNONYMS.*

Firft wings dark brown, with three waved ftripes of yellowifh colour acrofs the upper wings. Second wings ferruginous brown.

PHALÆNA PYRAMIDEA.  *Noctua* criftata, alis fufcis, ftrigis tribus undatis flavefcentibus repandis, pofticis ferrugineis. *Linn. Syft. Nat.* 2. 856. 181.—*Fab. Spec. Inf.* 2. 232. 119.—*Ent. Syft. I.* 3. *p.* 2. 98. 290.

*Phalæna* feticornis fpirilinguis, alis deflexis, fuperioribus fufcis, lineis tranfverfis undulatis nigris, inferioribus ferrugineis. *Geoff. Inf.* 2. 160. 99.

*Phalæna* media, alis longis anguftis, exterioribus linealis et areolis nigris, albis atro rubentibus tranfverfis pulcre depictis, interioribus obfcure rubris. *Raj. Inf.* 159. 9.

---

The fingular pyramidal protuberance on the pofterior part of the Caterpillar has furnifhed an excellent fpecific name for the perfect Infect. It is found on the Oak, Sallow, and Blackthorn in May ;

E 2

changes

changes to the pupa ſtate the firſt week in June; the Moth comes forth in July. The Caterpillar ſpins a fine white ſilken web between two or three leaves in the manner repreſented in the plate at Fig. 1. Fig. 2. the chryſalis, which the web envelopes. Phalæna pyramidea is not a common ſpecies in this country.

P L A T E

# PLATE CXCIV.

## CHRYSOMELA FASTUOSA.

### COLEOPTERA.

### GENERIC CHARACTER.

Antennæ like a necklace of beads, encreasing in bulk towards the ends. No margin round the elytra or thorax.

### SPECIFIC CHARACTER

AND

### SYNONYMS.

Oval, shining like gold, with three longitudinal stripes of blue on the shells.

CHRYSOMELA FASTUOSA; ovata aurea, coleoptris lineis tribus coeruleis. *Fab. Syst. Ent.* 101. 36.—*Spec. Inf.* 1. 124. 48.—*Linn. Syst. Nat.* 2. 588. 18.
*Chryfomela* viridis nitida, thorace antice excavato, fasciis elytrorum longitudinalibus coeruleis. *Geoff. Inf.* 1. 261. 11.
*Coccinella faftuofa. Scop. Ent. carn.* 232.

––––––––––––––––––––

This beautiful Insect is rarely taken in England : we sometimes receive it from Germany, where it is more common ; a variety of it is also a native of North America.

E 3                                    The

The natural fize is fhewn at Fig. 1. In fome fpecimens blue is the predominant colour, in others a fine bronze or golden hue; the fexes are diftinguifhed by the colours in many inftances.

PLATE

# PLATE CXCV.

## SPHINX ZONATA.

### RED-BELTED SPHINX.

### LEPIDOPTERA.

### GENERIC CHARACTER.

Antennæ thickeft in the middle. Wings when at reft deflexed, Fly flow, morning and evening only.

### SPECIFIC CHARACTER.

Wings tranfparent, veined, margined with a band, or ftreak of black. Abdomen bearded at the extremity, black; with one fegment in the middle, red.

---

This is clearly an undefcribed Infeɛt. It bears the ftrongeft affinity to the *Sphinx Tipuliformis* of Linnæus; but as the zone or belt of red colour is an unerring diftinɛtion of our Infeɛt, it cannot belong to that fpecies. Fabricius having feparated the Linnæan fphinges into three new genera, *Sphinx*, *Sefia*, and *Zygaena*, this Infeɛt muft be included under the genus *Sefia* of his fyftem; two of the fpecies he has defcribed under that head, *S. culiciformis* and *tnthrediniformis* bear fome refemblance to our *Sphinx Zonata*, but are certainly diftinɛt fpecies.

Sphinx Zonata is rare in England; the natural fize is fhewn at Fig. 1. of the annexed plate. The fine purple appearance of the body difappears in fpecimens that have been kept long in a cabinet.

# PLATE CXCVI.

## PHALÆNA WAUARIA.

### L, or GOOSEBERRY MOTH.

### LEPIDOPTERA.

### *GENERIC CHARACTER.*

Antennæ taper from the bafe. Wings in general deflexed when at reft. Fly by night.

### *SPECIFIC CHARACTER*

#### AND

### *SYNONYMS.*

Wings grey: four black, irregular ftripes on the interior part of the upper wings; one refembling letter L.

PHALÆNA WAUARIA: pectinicornis, alis cinereis, anticis fafciis quatuor nigris abbreviatis inæqualibus. *Linn. Syft. Nat.* 2. 863. 219.—*Fn. Sv.* 1248.—*Fab. Spec. Inf.* 2. 249. 43.

*Phalæna* minor, alis amplis cinereo albicantibus 4 in exteriorum margine maculis linearibus e rufo nigricantibus. *Raj. Inf.* 179.

*Merian Europ. I. tab.* 25. *fig.* 151.

*Frifch. Inf.* 3. *tab.* 3. *fig.* 1.

*Roef. Inf.* 1. *phal.* 3. *tab.* 4.

*Wilks Pap.* 52. *tab.* 2. *a.* 2.

*Ammiral. Inf. tab.* 13. *fig,* 2. 3.

Except

Except *Phalæna Groſſulariata* *, figured in the early part of this work, few ſpecies are more common than this, on the ſmaller kinds of fruit trees, but particularly the Gooſeberry.   Harris calls it the L Moth from a ſuppoſed reſemblance of that letter in ſome of the dark marks on the upper wings.

The young Caterpillars appear almoſt as early as the leaves, and change to chryſalis late in may; in this ſtate they remain about twenty days before the Moth is produced.

---

* Currant Moth.

PLATE

# PLATE CXCVII.

## SIREX GIGAS.

### LARGEST TAILED WASP.

### HYMENOPTERA.

## GENERIC CHARACTER.

Two ſtrong jaws. Palpi two. Antennæ filiform, of about twenty-four joints. Sting projected, ſerrated like a ſaw. Abdomen terminate in a ſpine. Wings lance-ſhaped.

## SPECIFIC CHARACTER

### AND

## SYNONYMS.

Abdomen of nine ſegments; the 3, 4, 5, 6, black; the others yellow. Thorax hairy.

SIREX GIGAS abdomine ferrugineo: ſegmentis. 3. 4. 5. 6 nigris, thorace villoſo. *Linn. Syſt. Nat.* 2. 928. 1.—
*Fn. Sv.* 1573.
*Fab. Ent. Syſt.* 2. *p.* 124. 139.
*Roeſ. Inſ.* 2. *Veſp. tab.* 8. 9.
*Sulz. Inſ. tab.* 18. *fig.* 114.
*Schæff. Icon tab.* 1. *fig.* 2. 3.
*Reaum. Inſ.* 6. *tab.* 31. *fig.* 1. 2.
*Degeer Inſ.* 1. *tab.* 36. *fig.* 1. 2.
*Seb. Muſ.* 4. *tab.* 53. *fig.* 15.

The

The Sirex genus, as it ftands in the *Entomologia Syftematica* of Fabricius, includes only twenty-fix fpecies; thefe are chiefly Euro-pean Infects; but very few are natives of this country. The Sirex Gigas is found in the north of Europe; it has been taken in England, but very rarely: Yeats and Berkenhout mention it as a Britifh fpecies, and we have been informed that it is fometimes taken in Scotland. It is likely to be met with in Pine forefts, as the female feems to prefer that wood to depofit her eggs in. As no Englifh Entomologift has attempted to defcribe the peculiar habits of this tribe of Infects, and efpecially of Sirex Gigas, the following particulars may be fatisfactory to our readers.

The extenfive forefts of Germany furnifhed the accurate Roefel with many opportunities of finding and obferving the metamorphofis of Infects that are rarely to be found in other parts of Europe; and this enabled him to favour the world with a particular defcription and feries of figures of all the changes of Sirex Gigas, in the *Bom-byliorum et Vefparum* of his *Infecten Beluftigung*. His figure of the female Infect agrees with that we have given; the male is confi-derably fmaller, and has no fting*. The fting of the female confifts of three parts, a fheath which divides into two parts or valves, and a fine inftrument fomewhat refembling a needle; it is with this in-ftrument it wounds its enemies, and the fting is faid to caufe an excruciating pain. The microfcope difcovers this part to be befet with a number of very minute teeth, like the edge of a faw: with this fting the creature can pierce the wood of found trees; for we fufpect, it does not always depofit its eggs in fuch as are decayed, but rather in fuch as will fupply the larva with nourifhment when it is hatched. The eggs are laid in clufters of two or three hundred together; they are of a pale yellow colour, about the thirtieth part of an inch in length, and fhaped like a weaver's fhuttle. The larva lives in the body of the tree, enlarging its habitation as it increafes in fize, for it never leaves the tree till it becomes a winged creature.

---

* This is a generical diftinction.

The

PLATE CXCVII. 41

The larva when full grown is about an inch and a quarter in length, and as thick as a goofe quill. It is a heavy fluggifh creature, almoft cylindrical, the head very fmall, and the whole of an uniform pellucid yellowifh colour. It has a fmall fpine at the end of the body like thofe by which the larva of fome fpinges are diftinguifhed: this fpine is alfo a ftriking charaĉter in the perfeĉt Sirex. In the pupa the form of the winged creature is more vifible than in the larva ftate; it is of a browner colour than the larva, and the rudiment of the fting and legs are very vifible.

In the early editions of the Syftema Naturæ of Linnæus, the firices are arranged with the ichneumons in one genus; but later obfervations induced Linnæus to make two genera of them in his laft works. *Geoffroy* and Schæffer have added fome particulars to the Linnæan generic charaĉter; thefe principally relate to the number of joints in the tarfi.

PLATE

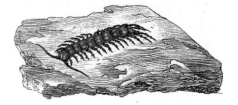

# PLATE CXCVIII.

## SCOLOPENDRA FORFICATA.

### APTERA.

#### WITHOUT WINGS.

### *GENERIC CHARACTER.*

The fame number of feet as fegments of the body. Antennæ fetaceous. Palpi two, jointed. Body depreffed or flat.

### *SPECIFIC CHARACTER*

#### AND

### *SYNONYMS.*

Feet fifteen on each fide.

SCOLOPENDRA FORFICATA: pedibus utrinque 15. *Linn. Syft. Nat.*
    2. 1062. 3.—*Fn. Sv.* 2064.—*Geoffr. Inf.* 2. 674. 1.
    —*Fab. Spec. Inf.* 2. 532. 3.—*Ent. Syft.* 2. 390. 4.
*Scolopendra* rufo fufca, pedibus utrinque 15. *Degeer Inf.* 7. 557. 1.
    *tab.* 35. *fig.* 12.
    *Sulz. Inf. tab.* 24. *fig.* 155.
    *Schæff. I. pl.* 46. 12.

---

This fingular creature not only changes its fkin feveral times; but, as it advances in age the fegments of which the body is compofed increafe in number: the Infect when full grown has nine fuch fegments. Linnæus mentions it as a native of America: in many

5

parts

parts of Europe it is not uncommon : in England it is found under loofe ftones in damp places, and runs fwiftly.

Fabricius has made a falfe quotation under this fpecies to the Fundamenta Entomologica of *Schæffer* ; as errors will occur in the moft accurrate works, we fhould not deem it neceffary to notice this circumftance, if he had not continued the fame miftake from his Species Infectorum publifhed in 1781 to his laft work Entomologia Syftematica, *emenda et aucta, &c.* publifhed in 1793.— his reference is *Schæff. Elem. tab.* 3. *fig.* 1.—On examining that part of Schæffer's works, we find the figure he quotes is a fpider ! Scolopendra forficata is given in the 46th plate of Vol. I. of that author's *Icones Infectorum circa Raftifbonam indigenorum, &c.* as quoted amongft the fynonyms above.

The Scolopendra forficata is, we believe, the largeft of the genus found in this country. In many parts of the world, fome kinds are found of a frightful fize and afpect ; the Scolopendra Morfitans of the Eaft-Indies, is about five inches in length, and as thick as a goofe quill. Sir G. Staunton, in his Hiftorical Account of the Embaffy to China, mentions the Scolopendras and fcorpions of that country : we have one fpecies of the former from China that exceeds in magnitude every one of the genus we have feen from other parts of the world, and is perhaps the largeft known ; it is near one foot in length, and is about one inch and an half round the girth of the body ; the colour is of a fine fhining cheftnut brown, the legs inclining to yellow. The moft fingular Infect of this kind in England is the Scolopendra electrica, which fometimes emits a fpark or flafh of light in the dark.

Fig. I. reprefents the underfide of the head and antennæ, magnified.

P L A T E

# PLATE CXCIX.

## PAPILIO C. *album*.

### COMMA BUTTERFLY.

### LEPIDOPTERA.

### *GENERIC CHARACTER.*

Antennæ terminate in a club. Wings, when at reft, erect. Fly by day.

### *SPECIFIC CHARACTER*

### AND

### *SYNONYMS.*

Wings deeply fcalloped, angulated, reddifh brown, with black fpots. Beneath, a white fpot refembling letter C on the under wings.

PAPILIO C. *album:* alis angulatis fulvis nigro maculatis, pofticis fubtus. C. albo notatis. *Lin. Syft. Nat.* 2. 778. 168.—*Fn. Sv.* 1059.—*Fab. Spec. Inf.* 2. 93. 409.
*Robert. Icon. tab.* 23.
*Merian. Europ. tab.* 14. *fig.* 1.
*Reaum. Inf.* 1. *tab.* 27. *fig.* 9. 10.
*Harris Aurel. tab.* 1. *fig. a—d.*
*Roef. Inf.* 1. *pap.* 1. *tab.* 5.
*Efp. pap.* 1. *tab.* 13. *fig.* 3.
*Bergftræff. Inf.* 2. *tab.* 38. *fig.* 1—5.
*Seb. Muf.* 4. *tab.* 1. *fig.* G.

---

In colours and markings the Comma Butterfly feems at firft fight allied to *Papilio Urticæ* (*Tortoifefhell Butterfly*), but the elegant fcal- lops of the margins of the wings fufficiently diftinguifhes it

F                                                              from

from every other Britifh fpecies: indeed wings indentated in this remarkable manner are rarely feen in any Infects, thofe from foreign countries not excepted. *Papilio Gaureum,* a native of Afia according to Fabricius, and figured amongft the Infects found in North America, by *Abbot,* approaches nearer to it than any diftinct fpecies we are acquainted with. The larva is known by a broad white fpace on the back that extends from the pofterior extremity to the eighth joint of the body: it has one peculiarity that muft not be omitted; when it is fufpended by the tail from a fmall twig or leaf, the head is bent upwards, and the whole creature perfectly refembles hook.

There is more than one brood of this fpecies in a year: the Butterflies are generally found in June, the fecond brood late in Auguft. The Caterpillars are fometimes found in February, but oftener in July; they remain about a fortnight in chryfalis.

PLATE

# PLATE CC.

## PAPILIO DAPLIDICE.

### BATH-WHITE BUTTERFLY.

### LEPIDOPTERA.

### *GENERIC CHARACTER.*

Antennæ terminate in a club. Wings erect when at reft. Fly by day.

### *SPECIFIC CHARACTER*

#### AND

### *SYNONYMS.*

Wings round, white. Margin brown, underfide marked with yellow, green, and white fpots.

PAPILIO DAPLIDICE : alis integris rotundatis albis, margine fufcis,
     fubtus luteo grifeis albo maculatis. *Lin. Syft. Nat.*
     2. 760. 81. *Papilio Daplidice. Pall Spicil. fafc.* 9.
     *tab.* 2.
     *Cramer Inf.* 15. *tab.* 171. *fig. C. D.*
     *Seb. Muf.* 4. *tab.* 23. *fig.* 11. 12. 15. 16.
     *Schæff. Icon. tab.* 79. *fig.* 2. 3.
     *Petiv. Gazoph. tab.* 1. *fig.* 7.

In a former defcription we have noticed the locality of one fpecies of the Butterfly tribe (*Pap. Hero*) in this country; of the Pap. Daplidice we have alfo to obferve, that it is only found in the envi-rons of Bath. We have been informed that it is found in fome

part

part of Scotland, but our information does not fully authorize us to give it as a native of that part of the British empire. If it is any where common, it is in Germany and in Hungary. It is found in Africa, and we have also a variety of it from North America. Fabricius does not mention it as a native of England; and it is so scarce with us that few English cabinets have a specimen of it. The trivial appellation *Bath-White* Butterfly has been given to it by some English Entomologists. The small number of foreign authors who have figured this species sufficiently notes its scarcity in Europe, and we are not acquainted with the works of any English author that has a figure of it.

P L A T E

# PLATE CCI.

## PANORPA COMMUNIS.

### Neuroptera.

### *GENERIC CHARACTER.*

Beak horny, cylindrical. Palpi 2.

### *SPECIFIC CHARACTER*

AND

### *SYNONYMS.*

Wings, equal length, fpotted with black.

PANORPA COMMUNIS: alis æqualibus nigro maculatis. *Lin. Syſt.*
  *Nat.* 2. 915. 1.
  *Fn. Sv.* 15. 16.
  *Geoff. Inf.* 2. 260. 1. *tab.* 14. *fig.* 2.
  *Degeer. Inf.* 2. 2. 86. *tab.* 24. *fig.* 3. 4.
  *Reaum. Inf.* 4. *tab.* 8. *fig.* 9.
  *Sulz. Hiſt. Inf. tab.* 25. *fig.* 5, 6.
  *Schæff. Icon. tab.* 88. *fig.* 7.

The Panorpa communis is very common in fields in the fummer. Its metamorphofis is entirely unknown.

The tail of the male is armed with a weapon, fomewhat refembling the fting of a fcorpion. The tail of the female has an appendage, but very different in its ftructure. Thofe parts of both fexes are reprefented in our plate, of the natural fize, and magnified.

G        PLATE

# PLATE CCII.

## FIG. I.

## PHALÆNA DITARIA.

### MAID OF HONOR MOTH*.

#### LEPIDOPTERA.

### GENERIC CHARACTER.

Antennæ taper from the bafe. Wings in general deflexed when at reft. Fly by night.

### SPECIFIC CHARACTER

#### AND

### SYNONYMS.

Antennæ feathered. Wings green, with brown fpots near the margins.

PHALÆNA DITARIA: pectinicornis alis viridibus: maculis marginalibus ferrugineis. *Fab. Ent. Syft.* 3. *p.* 2. 152. 82.
Phalæna bajularia: *Wien. Verz.* 97. 6.

---

This appears to be a local fpecies. It has been found for many years paft in the month of June, near the Oak of Honor, by Peckham. We have never heard that it has been taken in any other place in England.

---

* *Harris's Catal.* 31. 218.

Fabricius

Fabricius refers only to one author for this fpecies. He fays it feeds on the oak. We are uncertain whether it has ever been figured ;n any work before.

---

# FIG. II.

## PHALÆNA ALBICILLATA.

### CLIFDEN BEAUTY.

### LEPIDOPTERA.

### PHALÆNA.

### SPECIFIC CHARACTER

AND

### SYNONYMS.

Antennæ fetaceous. Wings whitifh in the difk, with a broad margin of pale black. A large dark fpot on the interior part of the upper wings.

PHALÆNA ALBICILLATA: feticornis alis omnibus nigricantibus: fafcia lata alba immaculata. *Linn. Syft. Nat.* 2. 870. 255.—*Fn. Sv.* 1278.—*Fab. Ent. Syft.* 3. 182. 190. *Clerk. Icon. tab.* 1. *fig.* 12. *Knoch. Beytr.* 1. 40. 1. *tab.* 2. *fig.* 8.

---

The trivial appellation, *Clifden Beauty*, has been given to this Infect by fome early Englifh entomologifts, who had not met with it in any other part of the country. It has fince been taken in feveral other parts of the kingdom; but in Kent particularly. Clerk deemed it worthy of a place among his rarities; and it is ftill confidered an exceeding fcarce and fine fpecies. Our fpecimen was taken in June, 1797, in Darent Wood, Dartford.

# PLATE CCII.

53

## FIG. III.

### PHALÆNA PORCELLATA.

CLIFDEN BEAUTY LIKENESS.

LEPIDOPTERA.

PHALÆNA.

### SPECIFIC CHARACTER

AND

### SYNONYMS.

Antennæ fetaceous. Firft wings white, with three brown bands, a white fpot on the exterior one. Second wings white, with a brown margin.

PHALÆNA PORCELLATA: feticornis alis anticis albis: fafciis tribus fufcus; media dimidiata, poftica marginali: macula alba. *Fab. Ent. Syft.* 3. 185. 201.
Phalæna Porcellata. *Wien. Verz.* 114. 1.

---

The fimilarity of this to the foregoing fpecies, induced us to figure both on the fame Plate. Phalæna Porcellata is very common in fummer. It appears in the winged ftate about the latter end of May. Linnæus has not defcribed this Infect. Fabricius has given its fpecific character, above quoted. We fufpect that it has not been figured by any author.

# 203

# PLATE CCIII.

## FULGORA EUROPÆA.

### EUROPEAN LANTERN CARRIER.

### HEMIPTERA.

## GENERIC CHARACTER.

Forehead elongated. Antennæ below the eyes, confift of two articulations. Beak or roftrum bent inwards under the body.

## SPECIFIC CHARACTER

### AND

## SYNONYMS.

Fore part of the head conic, body green. Wings tranfparent, reticulated.

FULGORA EUROPÆA: fronte conica, corpore viridi alis hyalinis reticulatis. *Linn. Syft. Nat.* 2. 704. **9.**
*Fab. Ent. Syft.* 4. *p.* 6. 19.
*Sulz. Hift. Inf. tab.* 9. *fig.* 5.
*Stoll. Cicad.* 1. *tab.* 11. *fig.* 51.

---

Simple as this little creature may appear to the general reader, the Entomologifts of this country efteem it a rarity; for it was not imagined that England produced any fpecies of its fingular genus till lately.

G 4

Mr.

Mr. Yeats, in his Inftitutions of Entomology, mentions two fpecies of it that have been caught in this country; one by Mr. *Hudfon*,
author of the *Flora Anglica*, the other by Mr. *Grey*. It is to be
lamented, that he does not inform us precifely what were the fpecies.
We learn that Fulgora Europæa was the Infeſt taken by Mr. Hudfon;
the other feems undetermined.

The Fulgora Europæa very much refembles fome of the Cicadæ
in form and fize, and have, therefore, been probably overlooked by
Englifh collectors of Infects. Fabricius defcribes it as a native of
France and Germany; but it is very fcarce in cabinets of foreign
Infects alfo. Perhaps it is not common in any country.

This fpecies does not exhibit any prominent features of its fingular genus; it has only a fmall conic hollow projection, or lantern,
on the fore-part of the head, inftead of the large projection that charac-
terife moft of the exotic Fulgoræ. It is not unlikely, however, that
it may poffefs, though in a fmall degree, the aftonifhing property of
diffufing a radiance of light, which particularly diftinguifh *F. Lan-
ternaria* of South America, *F. Candelaria* of China, and feveral
other fpecies. The light of fome of thefe Infects, according to the
reports of Naturalifts, and travellers in foreign countries, is fufficiently vivid and conftant to anfwer many purpofes. Some of the
Catholic miffionaries affert, that they could fee diftinctly to read and
write by the light of one of them; and that feveral of them being
faftened together, ferve to light the Indians when they travel in the
night.

The roftrum, or beak, through which the Fulgoræ Europæa (like
others of the fame genus) fucks its nutriment, feems to form a part
of the hollow projection in the front of the head; the tube lays clofe
to the belly, between the fix legs. To explain the fingular ftructure
of this tube or roftrum, we have given a front and a profile view
of it as it appeared under the lens of a microfcope.

4

Our

Our fpecimen is altogether green, except the wings; the recticu-lations of which are alfo of the fame colour; and as in Sulzer's figure, the clear parts of the wings were ftrongly tinged with green alfo.

PLATE

# 204

2

1

# P L A T E  CCIV.

## FIG. I.

### SPHINX LINEATA.

#### LEPIDOPTERA.

### *GENERIC CHARACTER.*

Antennæ thickeft in the middle.  Fly flow morning and evening only.

### *SPECIFIC CHARACTER*

#### AND

### *SYNONYMS.*

Firft wings greenifh, or olive colour, with bands and ftreaks of white.  Second wings black with a broad red band on each.

SPHINX LINEATA: alis virefcentibus: fafcia ftriifque albis, pofticis nigris; fafcia rubra.  *Fab. Ent. Syft. t.* 3. *p.* 1. 368. 39.
*Sphinx Daucus* Cram. Inf. 11. tab. 125. fig. D.
Sphinx Koechlini. *Fuefl. Arch.* 1. *tab.* 4.

––––––––––

This fine Infect is a native of Europe and America.  It has a place in every cabinet of Englifh Infects; but on what authority it will be difficult now to determine.  It is highly probable, that the teftimony of its difcovery in England is now forgotten, like that of Papilio Podalirius, figured in another part of this work.

The following defcription of its Caterpillar, which we have feen preferved, and in foreign drawings, will enable the curious in Infects to fearch after it with, at leaft, a diftant chance of fuccefs.  The

9                                                                   general

general colour of the Caterpillar is green, varied with yellow; and some streaks and spots of red down the back; it has also a large black spot on each side every segment; the head is black; and it has a spine, or tail. This is the appearance of it in one skin; it casts its skin several times, and will therefore vary in some degree from this account.—The pupa is yellowish brown, speckled with black. It feeds on *Ladies Bed-straw, Madder, Goose-grass*, &c.

-----

## FIG. II.

### PHALÆNA STATICES.

#### FORRESTER.

#### LEPIDOPTERA.

#### SPHINX.

### *SPECIFIC CHARACTER.*

First wings green blue; second brown.

SPHINX STATICES: *Linn. Syst. Nat.* 2. 808. 470.—*Fn. Sv.* 1098.
ZYGOENA STATICES: viridi cœrulea alis posticis fuscus. *Fab. Ent.*
    *Syst. T.* 3. *p.* 1.406. 68.
    *Geoff. Inf.* 2. 129. 40.
    *Robert. Icon. tab.* 30. *fig.* 1.
    *Petiv. Muf.* 35. 329.
    *Schæff. Icon. tab.* 1. *fig.* 9.
    *Efp. Inf.* 2. *tab.* 18. *fig.* 2.

-----

Found in the winged state in May.—Frequents meadows. The larva is described of a very deep black, with a line of white down the back, and some lunar spots of the same colour in different parts. It feeds on docks.

**PLATE**

# 205

# PLATE CCV.

## FIG. 1. 1.

### CURCULIO LAPATHI.

#### COLEOPTERA.

*GENERIC CHARACTER.*

Antennæ clavated, elbowed in the middle, and fixed in the fnout, which is prominent.

*SPECIFIC CHARACTER*

AND

*SYNONYMS.*

Snout long, two teeth on the thighs. White and black varied. Thorax and wing cafes rough with prickles.

CURCULIO LAPATHI: longiroftris femoribus bidendatis albo ni-
groque variis, thorace elytrifque muricatis.—*Linn.*
*Syft. Nat.* 608. 20.—*Fn. Sv.* 591.
*Fab. Ent. Syft.* 1. 429.
Curculio Lapathi: *Oliv. Inf.* 83. *fig.* 69. 6.
*Degeer Inf.* 5. 223. 16. *tab.* 7. *fig.* 1. 2.

―――――――――

Found on the Willow in May.

The figures 1. 1. exhibit the natural fize and magnified appear-ance.

FIG.

FIG.  2.  2.

CURCULIO  HORTULANUS.

Coleoptera.

Curculio.

*SPECIFIC  CHARACTER*

AND

*SYNONYMS.*

Nearly globular.   Afh colour, with two black fpots on the lon=
gitudinal future of the wing cafes.

Curculio Hortulanus: fubglobofus cinereus, punctis duobus
        nigro futuræ longitudinalis coleoptrorum.—*Geoff.* 1.
        298. 48.
        *Villers.* 1. 202. 118.

———————————

Fabricius has omitted this fpecies in his works, though Geoffroy
and Villers have both defcribed it.   It is found on plants of the
fcrophularia genus (figwort.)

FIG.

PLATE CCV. 63

# FIG. 3. 3.

## CURCULIO AVELLANÆ.

COLEOPTERA.

CURCULIO.

*SPECIFIC CHARACTER.*

Black. A lunated, oblique, whitifh mark near the bafe, and a white fpot near the apex of each wing cafe.

CURCULIO AVELLANÆ: nigra elytris bafi interne lunula fafciaque ante apicem albis.

———————————

This minute Infect feems to form an intermediate fpecies between *Salicis* and *C. Caprea.* It is an undefcribed Infect. In the MS. of T. Marfham, efq; it ftands under the fpecific name Avellanæ.— Was found on the Willow in June.

PLATE

# 206

# PLATE CCVI.

## PAPILIO IO.

### PEACOCK BUTTERFLY.

### LEPIDOPTERA.

## GENERIC CHARACTER.

Antennæ clubbed at the end. Wings erect when at reft. Fly by day.

## SPECIFIC CHARACTER

### AND

## SYNONYMS.

Wings angulated, indented. Bright brown, with fpots of black. A large blue eye on each wing.

PAPILIO IO: alis angulato dentatis fulvis nigro maculatis: fingulis ocello coeruleo.—*Linn. Syft. Nat.* 2. 769. 131.—*Fn. Sv.* 1048.—*Fab. Ent. Syft. I.* 3. *p.* 4. 88. 276.
*Roef. Inf.* 1. *pap.* 1. *tab.* 3.
*Wilk. Pap. tab.* 3. *a* 2.
*Reaum. Inf.* 1. *tab.* 25. *fig.* 1. 2.
*Schæff. Icon. tab.* 94. *fig.* 1.
*Merian. Europ.* 1. *tab.* 26.
*Albin. Inf. tab.* 3. *fig.* 4.
*Goed. tab.* 1. *fig.* 23.
*Efp. Pap.* 1. *tab.* 5. *fig.* 2.

H                                        We

We have not a more beautiful Insect in this country than the Peacock Butterfly. It is, indeed, too common to claim the particular notice of Entomologists; but to those who admire most the splendid species of this beautiful tribe of creatures, it will probably prove acceptable. The underside is entirely of a shining dark colour, with innumerable waved streaks of black. The upperside is represented in the annexed plate.

The Caterpillars, which are black, beset with spines, and elegantly marked with rows of white spots, are frequently found feeding on the nettles, and other low herbage by the sides of ditches. They change to the chrysalis state the first week in July, and appear in the winged state twenty-one days after.

P L A T E

1

# PLATE CCVII.

## JULUS TERRESTRIS.

### APTERA.

### GENERIC CHARACTER.

Feet on each fide double the number of the fegments of the body, Antennæ beaded Palpi 2. jointed. Body femicircular.

### SPECIFIC CHARACTER

### AND

### SYNONYMS.

Feet 200.

JULUS TERRESTRIS: pedibus utrinque 100. *Linn. Syft. Nat.* 2. 1065. 3.—*Fn. Sv.* 2066.—*Fab. Ent. Syft. I.* 2. 394. 8. *Degeer Inf.* 7. 578.
*Geoff. Inf.* 679. 1.
*Frifch. Inf.* 2. tab. 8. *fig.* 3.
*Sulz. Inf. tab.* 24. *fig.* 156.
*Sulz. Hift. Inf. tab.* 30. *fig.* 15.

---

This fingular creature is found of a vaft magnitude in foreign countries. We poffefs one of that kind between four and five inches in length. In Europe, or at leaft in England, they are feldom confiderably larger than the annexed figure.

It has two pair of feet to every fegment of the body. Thefe are very minute, but give a remarkable appearance to the Infect. It is found in damp places, generally under ftones.

PLATE

# PLATE CCVIII.

## PHALÆNA FIMBRIA.

### BROAD-BORDERED YELLOW-UNDERWING MOTH.

### LEPIDOPTERA.

### *GENERIC CHARACTER.*

Antennæ taper from the bafe. Wings in general deflexed when at reft. Fly by night.

### *SPECIFIC CHARACTER*

#### AND

### *SYNONYMS.*

#### NOCTUA.

Thorax crefted. Firft wings clay-colour, marked with obfcure bands or ftreaks. Second wings reddifh orange, with a broad bar of black.

PHALÆNA FIMBRIA: criftatata alis planis grifeo fafciatis; pofticis
   helvolis: macula lineari atra. *Linn. Syft. Nat.* 2.
   842. 123.—*Fab. Ent. Syft. T.* 3. *p.* 2. 59. 165.
   *Wien. Verz.* 87. 18.
   *Schreb. Inf. fig.* 9.

---

This rare Infect is diftinguifhed from two very common fpecies that are allied to it, by the broad border of black on the under wings, as its trivial name implies. In the larva ftate, it is one of that kind collectors denominate under-ground feeders: fubfifting chiefly on the roots of grafs, and never coming out of the ground till the evening, for which reafon it is very rarely taken.

I

The

The Moth is very delicate in its appearance ; the Caterpillar quite plain. Our fpecimen changed to cryfalis early in May, and produced the moth in the middle of June.

Fabricius was not informed that it was a native of this country, as appears by his laft work, in which he defcribes it only as a native of Germany.

PLATE

# PLATE CCIX.

## CERAMBYX LINEATOCOLLIS.

### COLEOPTERA.

### *GENERIC CHARACTER.*

Antennæ articulated, diminifhing in fize towards the end. Thorax gibbous, or fpined on the fides. Elytra narrow, and of equal breadth.

### *SPECIFIC CHARACTER.*

Entirely covered with hair, greenifh. Thorax unarmed with fpines, cylindrical, marked with yellow lines, fhells without fpots, brown.

CERAMBYX LINEATOCOLLIS: villofus viridefcens, thorace mutico cylindrico flavo-lineato, elytris immaculatis fufcis.— *Marfham's MS.*

---

We muft confider this as a new Britifh fpecies of Cerambyx, neither Linnæus nor Fabricius having given any defcription of it. It is defcribed only in the manufcripts of T. Marfham, Efq. whofe accurate definition of its characters we have adopted.

It appears to be a local fpecies. The only two fpecimens that have occurred to our notice, having been taken in the Ifle of Ely, Cambridgefhire. Our fpecimen was found on the bark of the willow.

# PLATE CCX.

## PHALÆNA LANESTRIS,

### LITTLE EGGER MOTH.

### LEPIDOPTERA,

## *GENERIC CHARACTER.*

Antennæ taper from the bafe. Wings in general deflexed when at reft. Fly by night.

### BOMBYX.

Antennæ of the male pectinated, of the female fetaceous.

## *SPECIFIC CHARACTER*

### AND

## *SYNONYMS.*

Wings ferruginous, firft pair with a white ftripe acrofs each: a white fpot near the bafe, and another in the middle of each.

PHALÆNA LANESTRIS: alis reverfis ferrugineis: ftriga alba, anticis
    puncto bafique albis.—*Linn. Syft. Nat.* 2. 815. 28.—
    *Fn. Sv.* 1105.
    *Fab. Ent. Syft.* 3. *p.* 1. 429. 68.
    *Wien. Verz.* 57. 2.
    *Roef. Inf.* 1. *phal.* 2. *tab.* 62.

---

The Caterpillars of the little Egger Moth, feed on black and white thorn, willow, lime-tree, &c. The female depofits a large clufter of eggs in a tuft of hair collected from her body. When

I 3

thefe

are hatched, the young begin to fpin a ftrong white web, which they enlarge as their fociety increafes; they remain together till they have devoured all the leaves of the plant on which they are hatched, or till they are arrived at full fize to change into the chryfalis ftate.

Thefe Caterpillars are not very uncommon in fome parts of the country, efpecially in Kent. They are ready to change to chryfalis ftate late in June. The Moth is not produced till April following.

The trivial Englifh name, Egger Moth, is given to this, and two or three other Moths, from the fimilitude of the cafe in which the chryfalis is inclofed to the fhape of egg.

PLATE

# 211

# PLATE CCXI.

## PAPILIO MACHAON.

### SWALLOW-TAIL BUTTERFLY.

### LEPIDOPTERA.

## GENERIC CHARACTER.

Antennæ clubbed at the end. Wings erect when at reft. Fly by day.

## SPECIFIC CHARACTER

### AND

## SYNONYMS.

Wings and tails of a yellow colour, with broad fpaces of brown marked with yellow lunar fpots. A reddifh fpot on the interior angle of the lower wings.

PAPILIO MACHAON: alis caudatis concoloribus flavis: limbo fufco; lunulis flavis, angulo ani fulvo.—*Linn. Syft. Nat.* 2. 750. 33.—*Fn. Sv.* 1031.—*Fab. Ent. Syft.* 2. *p.* 1. 87. *Roef. Inf.* 1. *pap.* 2. *tab.* 1. *Wilk. Pap. tab.* 47. *tab.* 1. *a* 1. *Merian. Europ. Inf.* 94. *Frifch. Inf.* 2. *tab.* 10. *Schæff. Icon. tab.* 45. *fig.* 1, 2. *Seba Muf.* 4. *tab.* 32. *fig.* 9, 10. *Geoff. Inf.* 2. 54. 23. *Efp. Pap.* 1. *tab.* 1. *fig.* 1.

Papilio

Papilio Machaon and Papilio Podalirius are the only two fpecies of Swallow-tail Butterflies that have been found in England. Both are very fcarce, but Papilio Machaon lefs fo than Papilio Podalirius, of which a figure has been given in another part of this work.

Entomologifts mention feveral parts of the country in which it has been taken, both in the larva and winged ftate. Harris fays it feeds on wild fennel and carrots; that one he found remained in the chryfalis ftate from the 23d of September to May the 15th following, and another, that changed July the 15th, produced a butterfly the 10th of Auguft. He adds, that the fpecies was found in the meadows of Briftol and Weftram.

From the number of foreign authors who have given figures of the Butterfly, we may imagine that it is very common on the Continent. Thofe preferved in cabinets of Englifh Infects are generally brought from Germany, from whence alfo we fometimes receive preferved fpecimens of the Caterpillars.

PLATE

# 212

# PLATE CCXII.

## MUTILLA EUROPÆA.

### EUROPEAN MUTILLA.

### HYMENOPTERA.

### *GENERIC CHARACTER.*

Generally want wings. Body covered with down. Thorax blunt at the bale. Sting pointed ; concealed within the body.

### *SPECIFIC CHARACTER*

### AND

### *SYNONYMS.*

Head black. Thorax red. Abdomen black ; margins of fome fegments whitifh.

MUTILLA EUROPÆA : nigra thorace rufo, abdominis fegmentis margine albo.—*Fab. Ent. Syft. t.* 3. 368. 9.

Mutilla Europæa, nigra abdomine fefciis duabus albis, thorace antice rufo.—*Linn. Syft. Nat.* 2. 966. 4.—*Fn. Sv.* 1727. *Sulz. Hift. Inf. tab.* 27. *fig.* 23, 24.

Apis Aptera : *Udm. Diff.* 98. *tab.* 17.

---

The Mutillæ feems lefs clearly defined than moft of the Linnæan genera. That author defcribed only a fmall number of the fpecies, and was even doubtful whether feveral that were placed under that divifion of his fyftem did not more properly belong to fome other, efpecially to the ichneumons, among which feveral apterous infects are included.

The

The caufe of this uncertainty may be partly attributed to our entire ignorance of their manner of life or transformations. Some of the Mutillæ have wings, and others are without. Authors have confidered the apterous Infects as the females, and the winged kind as the males, which opinion is countenanced by numberlefs inftances in almoft every clafs of Infects. Others have however maintained that both males and females were winged, and that the apterous Infects were neuters, prefuming in fupport of fuch opinion, that the Mutillæ lived in focieties like the Wafps, Ants, and Bees.—From obfervations on a number of exotic fpecies of this tribe, we have no doubt that the winged Infects are males, and the apterous kind females.

Yeats alludes to three fpecies of Mutillæ that have been found in England, but names only the Mutillæ Europæa; and this is the only kind we have ever found. We have taken it on a fandy pathway, near the entrance of Coombe Wood, Surry.

P L A T E

# 213

# PLATE CCXIII.

## PHALÆNA PRÆCOX.

### LEPIDOPTERA.

### *GENERIC CHARACTER.*

Antennæ taper from the bafe. Wings in general deflexed when at reft. Fly by night.

### *SPECIFIC CHARACTER*

AND

### *SYNONYMS.*

### NOCTUA.

Thorax crefted. Wings deflexed. Firft pair afh-colour with two fpots on each, and a fhort dafh of red near the ends. Second pair reddifh brown.

PHALÆNA PRÆCOX : criftata alis deflexis cinereis bimaculatis: pofticis fafcia rufa abbreviata.—*Linn. Syft. Nat.* 2. 854. 174.—*Fab. Ent. Syft. I.* 3. *p.* 2. 97. 289. *Roef. Inf.* 1. *phal.* 2. *tab.* 51.

When the late Duchefs of Portland honoured the fcientific as well as practical part of Entomology with her patronage, her Grace difcovered, and reared from the caterpillars feveral fpecies of Phalæna, of which collectors were ignorant before. Phalæna Præcox is among the number of thofe her Grace found in one of the Portland ifles; and the fpecimen we have figured is one which formed part of her collection.

Fabricius

Fabricius fays the larva feeds on the thiftle. The rarity of this creature induced us to depart from our ufual method, and copy the larva and pupa from the plate in the works of Roefel, apprehending it would be particularly interefting to Englifh naturalifts in general, as that author alone has reprefented it in thofe ftates; and no collector that we are informed has met with it within the laft fifteen years.

P L A T E

# 214

# PLATE CCXIV.

## PHALÆNA RUSSULA.

### Clouded Buff Moth.

### Lepidoptera.

### *GENERIC CHARACTER.*

Antennæ taper from the bafe. Wings in general deflexed when at reft. Fly by night.

### *SPECIFIC CHARACTER*

### AND

### *SYNONYMS.*

Wings deflexed, bright yellow. Margin and antennæ blood red. A lunar-fhaped fpot on the middle of the wings.

Phalæna Russula : alis deflexis luteis: margine fanguineo lunu-
laque fufca, antennis fanguineis.—*Fab. Ent. Syfl. I.* 3.
*p.* 1. 180.—*Linn. Syfl. Nat.* 2. 830. 71.
*Schæff. Icon. tab.* 83. *fig* 4, 5.
*Clerk. Icon. tab.* 4. *fig.* 1.
*Raj. Inf.* 228. 75.

---

As the Phalænæ are not remarkable for a variety of gay colours, like thofe of the Papilio genus, an exception to a general rule in the beautiful fpecies before us, more ftrongly demands our notice. The male Phalæna Ruffula, which is known by the pectinated antennæ, is of a fine golden yellow, with a rich, though narrow marginal band of fanguineous red round the wings. The female is a pretty Infect, but is more inclined to brown throughout than the male.

This

This species has been suppofed to feed on grafs in the larva ftate, but as collectors have very rarely reared it from that ftate, it has been difficult to determine its proper food. Fabricius mentions lettuce and fcabious or devil's-bit. The larva is hairy, and in many refpects very much refembles that of the Garden Tiger Moth, from which we may perhaps infer that it is a general feeder.

We found the larva in May; fhortly after it fpun a web and paffed into the pupa ftate, from which the moth was produced the 11th of June following.

PLATE

# 215

# PLATE CCXV.

## PHALANGIUM CRANCROIDES.

### APTERA.

### GENERIC CHARACTER.

Eight feet. Four eyes, two on the fummit of the head, and two others on the fides. Antennæ refemble feet, and are placed at the fore-part of the head. Abdomen round.

### SPECIFIC CHARACTER

#### AND

### SYNONYMS.

Body of an oblong ovated form, flat. Claws fmooth, hairy at at the ends.

PHALANGIUM CANCROIDES: abdomine obovato depreffo, chelis lævibus: digitis pilofis.—*Lin. Syft. Nat.* 2. 1028. 4. —*Fn. Sv.* 1968.

SCORPIO CANCROIDES: abdomine ecaudato ovato depreffo fufco, manibus oblongis.—*Fab. Ent. Syft. T.* 2. 436. 10.

Chelifer abdomine lineis tranfverfis.—*Geoff. Inf.* 2. 618. 1.

Chelifer europæus obfcure fufcis corpore ovato depreffo, chelis elon-gatis.—*Degeer Inf.* 7. 355. 2. *tab.* 9. *fig.* 4.

*Roef. Inf.* 3. *tab.* 64.

*Frifch. Inf.* 8. *tab.* 1.

*Schæff. Elem. tab.* 38.

The fynonyms fufficiently denote the unfettled opinion of eminent naturalifts in refpect of the proper genus to which our Infect fhould be referred. We have followed the definition of Linnæus, becaufe

it

it appears to us more characteristic of the creature, which should have a lengthened articulated tail, terminated in a sharp crooked sting, to warrant us in placing it among the Scorpions. In the system of Fabricius this forms no part of his generical character, but those who are accustomed to depend only on the writings of Linnæus, would be perplexed to reconcile the apparent difference between the Phalangia of that author, and the Scorpio of Fabricius.

The general appearance of this creature, except the want of tail, is precisely that of a Scorpion in miniature.

Mr. Adams, in his Essay on the Microscope, has figured and described a new species of this genus; it is smaller, and differs in form from our present species : he calls it the Lobster Insect. We believe Phalangium Cancroides is the largest Insect of the genus found in England that resembles a Scorpion.

This Insect is sometimes found in the covers of old books, in rotten wood, and other damp and decayed substances. We once found it fastened on the body of the *Musca Vomitoria*, Common Flesh Fly, from which it could not be extricated without killing and tearing the fly into pieces.

Roesel has given a figure of it, and represented a parcel of its eggs. They are of an oblong form, colour greenish, and appear to be deposited in roundish clusters of about thirty or forty eggs in each.

The natural size of our Insect is represented at Fig 1.

P L A T E

# PLATE CCXVI.

## PHALÆNA LIBATRIX.

### HERALD MOTH.

### LEPIDOPTERA.

### *GENERIC CHARACTER.*

Antennæ taper from the bafe.  Wings in general deflexed when at reft.  Fly by night.

### *SPECIFIC CHARACTER*

AND

### *SYNONYMS.*

Thorax crefted.  Wings deflexed, varied with red and grey; two white fpots on the anterior wings; edges deeply ferrated or indented.

PHALÆNA LIBATRIX: criftata alis incumbentibus dentato erofis rufo grifeis: punctis duobus albis.—*Lin. fyft. Nat.* 2. 831. 78.—*Fn. Sv.* 1143.
*Fab. Ent. Syft.* L. 3. *p.* 2. 64. 181.
*Wien. Verz.* 62. 1.
*Geoff. Inf.* 1. 121. 26.
*Goed. Inf.* 1. tab. 67.
*Albin. Inf.* tab. 32. *fig.* 50.
*Schæff. Icon.* tab. 24. *fig.* 1. 2.
*Roef. Inf.* 4. tab. 20.
*Harris Inf.* tab. 1. *fig. C. D.*
*Pod. Inf.* 92. tab. 2. *fig.* 9.

K

The

The Caterpillar of this Infect is generally found under the bark of the willow and fallow, or on the rofe. It is probable there are two broods of it in the year, being fometimes taken in the winged ftate early in the fummer, but more commonly in the month of October; this is the more likely, as the Englifh Aurelians firft called it the Herald, from an idea that its appearance warned them of approaching winter.

This Infect remains about twenty-eight days in the pupa ftate, the Caterpillar not being found till the beginning of September.

# LINNÆAN INDEX.

## TO

## VOL. VI.

---

## COLEOPTERA.

---

## HEMIPTERA.

---

## LEPIDOPTERA.

L                                        Papilio

# I N D E X.

## N E U R O P T E R A.

HYMENOP-

# INDEX.

## HYMENOPTERA.

----

## APTERA.

L 2

ALPHA-

# ALPHABETICAL INDEX

TO

# VOL. VI.

# I N D E X.

ERRATA.

# ERRATA.

Page 23, For *Phalæna Menthraſtiri*, read *Phalæna Menthraſtri*.—Second line, deſcrip. pl. 189.

Page 60, —— Phalæna ſtatices, *read* Sphinx ſtatices. —Second line, deſcrip. pl. 204. fig. 2.

# THE

# NATURAL HISTORY

OF

# BRITISH INSECTS;

EXPLAINING THEM

IN THEIR SEVERAL STATES,

WITH THE PERIODS OF THEIR TRANSFORMATIONS,
THEIR FOOD, OECONOMY, &c.

TOGETHER WITH THE

## HISTORY OF SUCH MINUTE INSECTS

AS REQUIRE INVESTIGATION BY THE MICROSCOPE.

THE WHOLE ILLUSTRATED BY

# COLOURED FIGURES,

DESIGNED AND EXECUTED FROM LIVING SPECIMENS.

———————

By E. DONOVAN.

———————

VOL. VII.

———————

LONDON:

PRINTED FOR THE AUTHOR,

And for F. and C. RIVINGTON, Nº 62, ST. PAUL'S CHURCH-YARD.

MDCCXCVIII.

5

217

# THE

# NATURAL HISTORY

OF

# BRITISH INSECTS.

---

## PLATE CCXVII.

### PAPILIO HIPPOTHOE.

#### GREAT COPPER BUTTERFLY.

#### LEPIDOPTERA.

### GENERIC CHARACTER.

Antennæ clubbed. Wings erect when at rest. Fly by day.

### SPECIFIC CHARACTER

AND

### SYNONYMS.

Wings intire, margin white. Underside afh colour, with numerous black eye-fhaped fpots.

PAPILIO HIPPOTHOE alis integris: margine albo, fubtus cinereis:
punctis ocellaribus numerofis. *Linn. Syft. Nat.* 2.
793. 254.

B 2                                              *Fab.*

*Fab. Spec. Inf.—Ent. Syft.* 2. *T.* 3. *p.* 1. 309. 172,
*Degeer Inf.* 2. *tab.* 2.
*Roef. Inf.* 3. *tab.* 37. *fig.* 6. 7.
*Efp. pap. tab.* 38. *fig.* 1.
*Ernft, Inf. Europ.* 1. *tab.* 44. *fig.* 92. 93.

Papilio Hippothoe is the largeft and rareft of that kind of Butter-flies called *Coppers,* by Englifh collectors of Infects. We have not heard that it has been taken in this country for fome years paft : our fpecimens were met with in Scotland.

The female is larger than the male ; it has alfo a greater number of black fpots on the wings.

PLATE

# [ 5 ]

# PLATE CCXVIII.

## FIG. I.

### CIMEX GONYMELAS.

BLACK-KNEE FIELD BUG.

HEMIPTERA.

*GENERIC CHARACTER.*

Roſtrum infleĉted. Antennæ longer than the thorax. Back flat. Thorax margined.

*SPECIFIC CHARACTER.*

Brown. Abdomen red. Antennæ annulated with black. Knees of the ſame colour.

---

We conſider this as a nondeſcript Inſeĉt. It was taken at Darent Wood, Kent, early in May.

---

## FIG. II.

### CIMEX HAEMORRHOIDALIS.

HEMIPTERA.

CIMEX.

*SPECIFIC CHARACTER*

AND

*SYNONYMS.*

Greeniſh. Spines of the Thorax obtuſe. Breaſt-piece terminate in a long ſpine. Antennæ black.

CIMEX

CIMEX HAEMORRHOIDALIS: thorace obtuſe, ſpinoſo ſubvireſcens,
antennis nigris, ſterno porrecto. *Linn. Syſt. Nat.*—
*Fn. Sv.—Fab. Ent. Syſt.* 4. *p.* 98. 76.

---

This Inſect was found at the ſame time and place as the preceding ſpecies. It is the moſt elegantly coloured creature of its
tribe we have hitherto found. *Cimex Luridus* is more beautiful in
the larva, but not in the winged ſtate.

P L A T E

# 219

# PLATE CCXIX.

## PHALÆNA PRODROMARIA.

### OAK-BEAUTY MOTH.

### LEPIDOPTERA.

### GENERIC CHARACTER.

Antennæ setaceous. Wings in general deflexed, when at rest. Fly by night.

### SPECIFIC CHARACTER

### AND

### SYNONYMS.

Antennæ feathered. Wings white, speckled with numerous black spots. Two irregular, and nearly transverse bars of dark brown, on the upper wings.

PHALÆNA PRODROMARIA, pectinicornis alis albis nigro punctatis; fasciis duabus latis fuscis. *Fab. Ent. Syst. T.* 3. *p.* 1. 159. 105.

Phalæna Prodromaria. *Wien. Verz.* 99. 1.

---

The larva of this Moth, like others of the *geometræ*, raises itself when walking, into the form of an arch or loop : it is of an obscure grey and brown colour, faintly mottled : the head is red. This larva is seldom taken, and when taken, is reared to the fly state with the utmost difficulty. It seems a local species ; for we

have

have never heard that it has been found, except on the Oak trees *
in *Richmond Park*. It feeds on the higheft branches of the trees,
but defcends into the earth to become a pupa. It appears in the
fly ftate in March.

The male Infect is confiderably fmaller than the female. Its
horns, or antennæ, are alfo larger, and more feathered. This is a
fcarce Infect. It is found in Germany; and a variety of it has
been received from North America.

---

* It feeds alfo on Lime trees.

PLATE

# 220

# PLATE CCXX.

## PHRYGANEA RHOMBICA.

### SPRING FLY,

### NEUROPTERA.

### *GENERIC CHARACTER.*

Mouth furniſhed with four palpi. Antennæ longer than the Thorax. Firſt Wings lay horizontally on the body. Under Wings folded, and concealed beneath.

### *SPECIFIC CHARACTER*

#### AND

### *SYNONYMS.*

Wings greyiſh brown. Firſt pair marked with rhombic whitiſh ſpots.

PHRYGANEA RHOMBICA alis griſeis: macula laterali rhombica albæ.
*Linn. Syſt. Nat.—Fab. Ent. Syſt. T. 4. p. 77. 13.*
*Roeſ. Inſ. 2. Aqu. 2. tab. 16.*
*Schæff. Icon. tab. 99. fig. 5. 6.*

---

The Phryganea undergo their transformations in the water: in the larva ſtate they are taken by the fiſhermen for bait; and, in ſome parts of Holland, are found ſo abundant, that they are uſed as a cheap manure for the land. In the larva ſtate, they gene-

C          rally

rally form a fort of covering, or tube, for their defencelefs bodies. It is open only at one end, at which its head and fore legs are protruded, to take its prey. Some fpecies form thefe coverings of weeds and fmall fhells, gravel, fand, &c. That of our prefent fpecies, is compofed of little pieces of the ftalks of grafs, cut into an even form, and laid tranfverfely on each other. It attaches this tube to the roots of fome aquatic plants, and undergoes its transformations in it. In the annexed plate, we have reprefented the larva taken from the tube, and the pupa having the tube opened to exhibit its fituation therein.

The Fly is very common about ponds, rivers, and marfhy places.

P L A T E

# PLATE CCXXI.

## PHALÆNA MYRTILLI.

### SCARCE BROAD BORDER YELLOW UNDERWING MOTH.

#### LEPIDOPTERA.

### *GENERIC CHARACTER.*

Antennæ taper from the bafe. Wings in general deflexed when at reft. Fly by night.

### *SPECIFIC CHARACTER*

#### AND

### *SYNONYMS.*

Thorax crefted. Wings deflexed, brown, fpotted with white. Anterior wings yellow, with a deep black border.

PHALÆNA MYRTILLI criftata alis deflexis ferrugineis albo maculatis: pofticis luteis, fafcia lata fubmarginali nigra. *Lin. Syft. Nat.*—*Fab. Ent. Syft. T. 3. p. 2. 126. 379.*

---

A fmall Infect, but of fingular beauty ; it feeds on the whortle berry and floe.

This fpecies has been taken by Mr. Crow, of Faverfham. The only fpecimen we ever met with, was found in the caterpillar ftate, in Kent, in the month of May. The Fly came forth in June.

PLATE

# PLATE CCXXII.

## FIG. I.

### CARABUS VIOLACEUS.

#### COLEOPTERA.

*GENERIC CHARACTER.*

Antennæ fetaceous. Thorax fomewhat heart fhaped, margined. Elytra margined alfo.

*SPECIFIC CHARACTER*

**AND**

*SYNONYMS.*

Apterous, black, Margin of the Thorax and Wing cafes, gloffy violet. Edges fmooth.

CARABUS VIOLACEUS apterus niger thorace elytrorumque margi-
nibus violaceis, elytris lævibus.—*Fab. Ent. Syft.* I.
19. 125.
Carabus Violaceus. *Paykull Monogr.* 12. 4.
*Frifch Inf.* 13. *tab.* 23.

───────────

The larva of fome Carabi live in the ground, others in decayed wood. They prey on the fmaller kinds of Infects. Fabricius de-fcribes one hundred and ninety-five fpecies; a confiderable number of thefe are natives of Europe. Carabus Violaceus is found in fields.

D                                          FIG.

## FIG. II.

### CARABUS GEMMATUS.

COLEOPTERA.

CARABUS.

*SPECIFIC CHARACTER*

AND

*SYNONYMS.*

Apterous, black. Wing cafes marked with ftriæ; and three rows of indented double fpots, bronzed.

CARABUS GEMMATUS apterus niger elytris ftriatis: punctis æneis
      bilobis excavatis triplice ferie.—*Fab. Ent. Syft.* 1. 19.
      127.
Carabus ftriatus.—*Degeer Inf.* 4. 90. 5. *tab.* 3. *fig.* 1.
Carabus gemmatus.—*Paykull Monogr.* 15. 6.

———————————

This fpecies has commonly been miftaken for *Carabus hortenfis:* the difference, however, between the two Infects, is confiderable. The colour of the Beetle is black; but when not damaged, is entirely covered with a rich bronze, partaking of a green and golden hue on the wing cafes, and a fine purple on the thorax: the underfide is plain black.

FIG.

PLATE CCXXII.    15

# FIG. III.

## CARABUS GRANULATUS.

COLEOPTERA.

CARABUS.

*SPECIFIC CHARACTER*

AND

*SYNONYMS.*

Apterous, black, bronzed. Wing cafes ftriated; three rows of elevated, or convexed-oblong fpots, with an intermediate elevated line on each.

CARABUS GRANULATUS apterus nigricans elytris æneis ftriatis in-
    terieɛtis punɛtis elevatis longitudinalibus.—*Lin. Syʃt.*
    *Nat.*—*Fab. Ent. Syʃt.* 1. 130. 28.
Carabus granulatus.—*Paykull Monogr.* 19. 9.
    *Degeer Inʃ.* 4. 88. 2.
    *Sulz. Hiʃt. Inʃ. tab.* 7. *fig.* 2.
    *Schæff. Icon. tab.* 18. *fig.* 6. *& tab.* 15. 6. *fig.* 1.

—————

Some authors fay, this fpecies is very common in the fields near London. It is often found in Batterfea meadows; and we have not found it elfewhere.

PLATE

# PLATE CCXXIII.

## FIG. I.

### PHALÆNA DERASA.

BUFF ARCHES MOTH.

LEPIDOPTERA.

### GENERIC CHARACTER.

Antennæ taper from the bafe. Wings in general deflexed when at reft. Fly by night.

### SPECIFIC CHARACTER

AND

### SYNONYMS.

Crefted. Wings deflexed. Anterior pair buff colour, with fmall arched markings.

NOCTUA DERASA: criftata, alis deflexis, anticis fupra decorticatis.
    *Fab. Syft. Ent.* 609 80.—*Spec. Inf.* 2. 229. 103.—
    *Ent. Syft.* 3. *p.* 2. 85. 250.
Phalæna derafa. *Linn. Syft. Nat.* 2. 851. 158.
Phalæna pyritoides. *Naturf.* 2. *tab.* 1. *fig.* 7. (*mas*).
Borkhaufen, *enr. Schmett.* 4. *T. n.* 281. *p.* 657.
Die Himbeereule. Der Wifchflügel. *Panz. Faun. Inf. Germ.*

———

A rare fpecies, is found in the Fly ftate early in Auguft.

E                                                        The

The larva of this phalæna is unknown to us, and has neither been figured or defcribed in any of the entomological works recently publifhed. The notes of Harris are not altogether fatisfactory; he mentions the time of its changing from the caterpillar to the pupa, but has given no figure or defcription of either. The ento-mologifts of Germany, where the phalæna is not fcarce, feem unacquainted with its metamorphofis. Fabricius, the latest writer on the fubject, has not defcribed it.

---

# F I G. II.

## PHALÆNA TRAGOPOGINIS.

### GOAT'S-BEARD MOTH.

#### LEPIDOPTERA.

#### PHALÆNA.

*SPECIFIC CHARACTER*

AND

*SYNONYMS.*

Anterior wings dark brown, with three black points or fpots in the centre, pofterior pair livid.

NOCTUA TRAGOGINIS: crifta, alis deflexis, anticis fufcis, punc-tis nigris tribus approximatis, pofticis lividus. *Fab. Syft. Ent.* 615. 107.—*Spec. Inf.* 2. 237.—*Ent. Syft.* 3. *p.* 2. 112. 336.

Phalæna Tragopoginis. *Lin. Syft. Nat.* 2. 855. 177.—*Fn. Sv.* 1189.

*Phalæna*

# PLATE CCXXIII. 19

*Phalæna* antennis filiformibus, alis deflexis fufcis nitidis, punctis tribus centralibus nigris, capite flavo. *Degeer Inf. Verf. Germ.* 2. 1. 303. 10. *tab.* 7. *fig.* 15.

---

Found on the Goat's beard, Spinach, and Docks.—Our fpecimen was taken in June.

---

# FIG. III.

## PHALÆNA LICHENES,

### LIVER-WORT MOTH.

### LEPIDOPTERA.

### PHALÆNA.

## SPECIFIC CHARACTER

### AND

## SYNONYMS.

Thorax crefted. Anterior wings green, with black marks. Pofterior pair brown. Underfide brown.

NOCTUA LICHENES : criftata, alis deflexis : anticis viridibus, maculis variis atris, fubtus fufcis. *Fab. Syft. Ent.* 614, 102.—*Spec. Inf.* 2. 235. 127.—*Ent. Syft.* 3. *p.* 2. *p.* 104. 312.

Noctua glandifera. *Wien. Verz.* 70. 2.

We have found this ſpecies againſt walls on which the *Lichen fuſco-ater* was growing. The larva is ſuppoſed to feed on plants of that genus. One ſpecimen was found in October, another early in the ſpring, from which we conclude there muſt be two broods of them in the year.

**P L A T E**

224

# PLATE CCXXIV.

## PHALÆNA NUPTA.

### RED UNDERWING MOTH.

### LEPIDOPTERA.

### *GENERIC CHARACTER.*

Antennæ taper from the bafe. Wings in general deflexed when at reft. Fly by night.

### *SPECIFIC CHARACTER*

#### AND

### *SYNONYMS.*

Thorax crefted. Anterior wings greyifh, varied with brown. Pofterior pair red, with two broad black waves acrofs. Abdomen hoary above, white beneath.

PHALÆNA NUPTA criftata alis planis cinerafcentibus: pofticis rubris; fafciis nigris, abdomine cano fubtus albo. *Linn. Syft. Nat.* 2. 841. 119.
*Wilks pap.* 33. *tab.* 1. *a.* 1.
*Roef. Inf.* 1. *phal.* 2. *tab.* 15.

---

The larva of the Red Underwing Moth feeds on the willow: it is found in that ftate in June and July. The Fly appears in Auguft, after having remained in the pupa ftate about twenty-one days.

E 3                                    Collecto

Collectors of Englifh Infects enumerate near twenty fpecies of Phalæna under the trivial diftinctions of *yellow underwing*, *copper underwing*, *orange underwing*, *pink underwing*, &c. &c. Among thefe the moft confpicuous both for beauty and magnitude, are the *red underwing*, and *crimfon underwing*. The firft is by no means uncommon in the winged ftate. The latter is very rare, or at leaft a local fpecies: it is found in the larva ftate on the tops of the higheft oaks in Richmond Park, and was formerly found in fimilar fituations in Burnt Wood, Effex. We are not informed that it has been taken in any other part of this kingdom.

Thefe two fpecies have been confounded with a third fort that is found in fome parts of Europe, but does not, we have every reafon to conclude, inhabit this country. This is the *Noctua Pacta* of Linnæus and Fabricius. Linnæus himfelf, in the firft editions of the Syftema Natura, confidered the Red Underwing Moth, figured by Roefel, tab. 15, as the Phalæna Pacta, and adds it in his Synonyms; but it appears corrected in the later editions *.

After that time, Harris, in his Aurelian, and other works, called the *Red Underwing* Phalæna Pacta, and the *Crimfon Underwing* Phalæna Nupta. And Dr. Berkenhout, following Harris, or inattentive to the exprefs language of the author he tranflated, has made the fame error in his Synopfis of the Natural Hiftory of Great Britain †. Indeed, it may be doubted, whether any later Englifh work on Infects has detected the error, for, examining a little tract of Mr. Matthew Martin, of the Bath Society, publifhed in 1785, we find the Red Underwing called therein *Phalæna Pacta*.

To place our remarks in a clear point of view, we need only quote the defcriptions of Linnæus:—" Noctua Pacta Criftata alis grifefcentibus fubundatis: pofticis rubris; fafciis duabus nigris. *Abdomine fupra rubro.*" And again in the general defcription:

---

* Corrected after 1759.    † Not corrected in the laft edition.

" *Abdomen*

# PLATE CCXXIV. 23

" *Abdomen fupra rofeum.*" Without adverting to the other characteriftic marks, this proves that the Linnæan fpecies of *Pacta* cannot be the fame with that of the authors before quoted, becaufe in their fpecies the upper part of the abdomen is hoary, inclining to brown, and not red *. Their Phalæna Pacta can be no other than the Phalæna Nupta of Linnæus and Fabricius; the precife fpecies reprefented in our plate; of which Linnæus and Fabricius fay: " Habitat in Europæ Salice Vitellina;" and of the Phalæna Pacta and Sponfa, " Habitat in Europæ Quercu." The firft lives on willows, the two others on oaks.

We clofe our remarks with obferving, that the above quoted Englifh authors have been no lefs miftaken as to Phalæna Nupta, which they have made the Crimfon Underwing Moth. We have before expreffed our doubt whether Phalæna Pacta has ever been found in this country;—we add, that the Infect, known to Englifh collectors by the trivial name of Crimfon Underwing, is the *Phalæna Sponfa* of Linnæus and Fabricius †, and confequently not connected in the leaft with Phalæna Nupta.

The readers of the works of Harris, Berkenhout, &c. are requefted to *read*

> For Phalæna Nupta, Phalæna Sponfa, Crimfon Underwing Moth.
> For Phalæna Pacta, Phalæna Nupta, Willow Red Underwing Moth.

And finally, remove Phalæna Pacta from the lift of Britifh fpecies, till it is proved to be a native of this country.

---

* Berkenhout fays the abdomen is reddifh above; but by this he only encreafes the miftake; for his fpecies agrees in every other refpect with the Willow Moth, on whic plant he alfo fays it is found. Page 140. Vol. 1.

† Vide Entomologia Syftematica. Vol. III. p. 2; p, 53. 147.

E 4   PLA

# PLATE CCXXV.

## SIREX SPECTRUM,

### BLACK-BODIED TAILED WASP.

### HYMENOPTERA.

Wings four, membraneous in general. Tail of the females armed with a sting.

## GENERIC CHARACTER.

Mouth armed with strong jaws. Palpi two, truncated. Antennæ filiform, containing upwards of twenty-four articulations. Sting projected, strong, and serrated.

## SPECIFIC CHARACTER

### AND

## SYNONYMS.

Abdomen black. Thorax rather hairy, a yellow stripe on each side, next the base of the wing.

SIREX SPECTRUM: abdomine atro, thorace villoso, litura ante alas lutea. *Fab. Syst. Ent.* 3. 26.—*Spec. Inf.* 1. *p.* 419. 109. 6.
Sirex spectrum. *Lin. Syst. Nat.* 2. 929. 3.—*Fn. Sv.* 1574.—*Degeer. Inf.* 1. *tab.* 36. *fig.* 6.—*Schæff. Icon. tab.* 4. *fig.* 9. 10.

This species bears much resemblance to some Insects of the Ichneumon genus. We have found it among the leaves of the Horseradish in June.

All the *firices* are rare in England. Sirex fpectrum is an active and vigorous creature, and which cannot be taken without danger of its ftinging. The fting is fmall, and fine as a needle, but formed of fuch hard or horny fubftance, that it will pierce the finger to the bone.

PLATE

# 226

# PLATE CCXXVI.

### VESPA VULGARIS.

#### COMMON WASP,

#### HYMENOPTERA.

Wings 4, membranous in general. Tail of the female armed with a fting.

### GENERIC CHARACTER.

Mouth armed with jaws. The fting fharp-pointed and concealed within the abdomen. Body fmooth, without hair. The upper wings folded.

### SPECIFIC CHARACTER

#### AND

### SYNONYMS.

A yellow line on each fide of the thorax : four yellow fpots on the fcutellam, a black belt, and two black fpots on each fegment of the abdomen.

VESPA VULGARIS : thorace utrinque lineola interrupta, Scutello qua quadrimaculato, abdominis incifuris punctis nigris diftinctis. *Fab. Syft. Ent.* 364. 9.—*Spec. Inf.* 1. 460. 9.—*Lin. Syft. Nat.* 2. 949. 4.—*Fn. Sv.* 1671.

*Vefpa* nigra luteaque, antennis totis nigris. *Degeer Inf.* 2. 2. III. *tab.* 26. *fig.* 7.

*Vefpa* thorace lineolis trium parium differentium flauves centium. *Geoff. Inf.* 2. 369. 2.

*Schæff.*

*Schæff. Elem. tab.* 130.
—— *Icon. tab.* 35. *fig.* 4.
*Reaum Inf.* 6. *tab.* 12. *fig.* 7. 8.

Moſt kinds of Waſps live in ſocieties ; and, like the bees, conſtruct combs, in which they depoſit their eggs, and rear their young. Some ſpecies are ſolitary, and each individual forms a neſt for itſelf.

The common Waſp lives in ſocieties : they collect the juices of fruits, infects, &c. and make honey, but it is inferior to that of bees. The metamorphoſis of the waſps and bees are ſimilar.

P L A T E

# PLATE CCXXVII.

## PHALÆNA MONACHA.

### BLACK ARCHES MOTH.

#### LEPIDOPTERA.

*GENERIC CHARACTER.*

Antennæ taper from the bafe. Wings in general deflexed when at reft. Fly by night.

*SPECIFIC CHARACTER*

AND

*SYNONYMS.*

Wings deflexed, white, with black arches. Abdomen red.

PHALÆNA MONACHA: alis deflexis albis atro undatis, abdominis incifuris fanguineis. *Lin. Syft. Nat.* 2. 821. 43.— *Fn. Sv.* 1130.—*Fab. Ent. Syft. T.* 3. *p.* 1. 446. 119. *Wien. Verz.* 52. 5. *Wilks pap.* 19. *tab.* 3. *a.* 4. *Schæff. Icon. tab.* 68. *fig.* 2, 3.

We have in few inftances been able to prefent a fpecies of Phalæna, with all its metamorphofis, more deferving attention than the Black Arches Moth. It is uncommonly rare in the winged ftate, and its larva and pupa is, we prefume, unknown to the Englifh Entomologifts at this time. We imagine Harris met with, and bred

this

this Infe&, though he has not figured it : he fays it fed on the Oak,
that it changed into chryfalis June 9th, and appeared in the winged
ftate July 9th, which is very near the time of our fpecimen changing.

The larva is rather a general feeder; for though Harris mentions
only Oak, we found that it would not refufe the leaves of fruit-trees,
fuch as Apples, Pears, &c. ; it feeds alfo on the Willow and Sallow.
The female is larger than the male, and has antennæ like briftles.

**P L A T E**

228

# PLATE CCXXVIII.

## SPHINX CONVOLVULI.

### BIND-WEED HAWK MOTH.

### LEPIDOPTERA.

### GENERIC CHARACTER.

Antennæ thickeſt in the middle. Wings, when at reſt, deflexed. Fly ſlow, morning and evening only.

### SPECIFIC CHARACTER

### AND

### SYNONYMS.

Wings entire, clouded. Poſterior pair marked with zigzag tranſverſe bands. Abdomen belted with alternate marks of red, black and white.

SPHINX CONVOLVULI: alis integris nebuloſis: poſticis ſubfaſcia-
tis, abdomine cingulis rubris atris albiſque. *Linn.*
*Syſt. Nat.* 2. 798. 6.—*Fab. Ent. Syſt. I.* 2. *p.* 1.
374. 54.
*Geoff. In.* 2. 86. 9.
*Roeſ. Phal.* 1. *tab.* 7.
*Sepp. Inſ.* 3. 19. *tab.* 4.
*Merian. Europ.* 39. *tab.* 75. *fig.* 2.
*Cramer Inſ.* 19. *tab.* 225. *fig. D.*
*Welks pap.* 10. *tab.* 1. 6. 2.
*Eſp. Inſ.* 2. *tab.* 5.
*Drury Inſ.* 1. *tab.* 25. *fig.* 4.

This

This is the largeft of the Hawk Moths that inhabits Great-Britain, except *Sphinx liguftri* and *Sphinx Atropos*. It is rarely taken in this country ; the curious in Englifh Infects have them from Germany, where they are more common than with us.

A beautiful variety of this Infect is found in North-America: the wings are more richly varied with different fhades of bright browns than the European kind ; the pofterior wings are of a fine rofe-colour. It has all the charaCteriftic marks of Sphinx Convolvuli, or we fhould hefitate to admit it as the fame fpecies. We received it from Mr. Abbot, in whofe folio work it is alfo figured ; he found it on the Wild Vine. Mr. Drury had the fame variety fent to him from St. Chriftopher's.

PLATE

# PLATE CCXXIX.

THE

## LARVA AND PUPA

OF

### SPHINX CONVOLVULI,

OR

#### BIND WEED HAWK MOTH.

———————

After much refearch, we have not been fo fortunate as to meet with the Larva of this rare Infect ; nor can we learn that it has been taken by any Collector of Englifh Infects for many years. In the winged ftate one Specimen was faid to be taken in the fields near Hoxton about two years ago.

To perfect the Hiftory of this fpecies, we have copied the Figures of the Caterpillar and Pupa, from N° 7, *Der Nacht-Voegel*, &c. &c. of Roefel's *Infecten Belluftigung*, Vol. I.

Our readers will obferve that the Fly produced from the Caterpillars reprefented by Roefel is nearly one-third larger than the fpecimens fuppofed to be bred in England; the Caterpillars muft therefore be larger in the fame proportion in the fpecimens found in Germany.—The Caterpillars are of two colours, one green with ftripes of yellow and fpots of black ; The other dull brown with ochre coloured ftripes, and fides of the fame. The Caterpillar figured by Abbot has a rofe-coloured band on the fide.

F                    PLATE

230

# PLATE CCXXX.

## FIG. I.

## PHALÆNA MAURA.

### OLD LADY MOTH.

#### LEPIDOPTERA.

### *GENERIC CHARACTER.*

Antennæ taper from the bafe. Wings in general deflexed when at reft. Fly by night.

### *SPECIFIC CHARACTER*

#### AND

### *SYNONYMS.*

Thorax crefted, Wings incumbent, exterior margins dentated, afh-colour, varied with large fpaces of black. On the underfide a deep whitifh border.

NOCTUA MAURA criftata, alis incumbentibus dentatis, cinereo nigroque variis, fubtus margine albo. *Fab. Syft. Ent.* 604. 61.—*Spec. Inf.* 2. 224. 81.—*Ent. Syft.* 3. p. 2. 63. 174.

*Phalæna maura* fpirilinguis criftata, alis depreffis dentatis, fafciis duabus nigris, inferioribus nigris, fafcia alba. *Linn. Syft. Nat.* 2. 843. 124.

Phalæna Lemur *Naturf.* 6. *tab.* 5. *fig.* 1. *Shœff. Icon. tab.* 1. *fig.* 5. 6.

---

This grave Moth appears in the month of Auguft: it frequents old houfes in evenings.—From its dingy appearance it is ufually called the Old Lady.

F 2

FIG.

## F I G. II.

### PHALÆNA LUCIPARA.

SCARCE ANGLE SHADES MOTH.

LEPIDOPTERA.

PHALÆNA.

*SPECIFIC CHARACTER*

AND

*S Y N O N Y M S.*

Crefted. Wings deflexed, greyifh, with angular dark marks.
An angular light coloured fpace on the exterior part of the Wing,
and a pale band acrofs the middle of each.

NOCTUA LUCIPARA criftata, alis deflexis cinereo nitidis, fafcia
media lata fufca. *Fab. Spe. Inf.* 2. 233. 121.——
*Ent. Syft.* 3. *p.* 2. *p.* 99. 244.

*Phalæna lucipara* fpirilinguis criftata, alis purpurafcentibus lucidis,
fafcia nigra, ftigmate poftico flavo. *Linn. Syft.*
*Nat.* 2. 857. 187.—*Fn. Sv.* 1201.

––––––––––

The common Angle-fhades Moth, *(Phalæna Meticulofa)* is figured
in a former part of this Work. Phalæna Lucipara is an Infect
nearly allied to it, but is far more fcarce ; we have only met with
the Specimen figured in the annexed Plate.—The Larva is fuppofed
to feed on the internal fubftance of Willows. The Fly has alfo
been obferved among thofe trees.

P L A T E

# PLATE CCXXXI.

## FIG. I. I.

### ATTELABUS APIARIUS.

#### COLEOPTERA.

*GENERIC CHARACTER.*

Antennæ thickeft towards the apex. Head protruded, broad, tapering towards the thorax. Four joints in each foot.

*SPECIFIC CHARACTER*

AND

*SYNONYMS.*

Bright blue : rather hairy. Wing cafes red, with three bars of blue : the third, at the termination of the apex.

ATTELABUS APIARIUS : *Lin. Syft. Nat.* 2. 620. 10.

CLERUS APIARIUS : fubnudus cyaneus elytris rubris : fafciis tribus
cœrulefcentibus : tertia terminali.—*Geoff. Inf.* 1.
304. 1. *tab.* 5. *fig.* 4.—*Fab. Ent. Syft.* 1. 209. 14.
*Degeer. Inf.* 5. 157. 1. *tab.* 5. *fig.* 3.
*Sulz. Inf. tab.* 4. *fig.* 6.

———————

A very local fpecies : we learn that it has been found near Manchefter.

F 3                    FIG.

## F I G. II.

### ATTELABUS FORMICARIUS.

*SPECIFIC CHARACTER*

AND

*SYNONYMS.*

Black. Thorax red. Wing cafes, with two bars of white: bafe red.

ATTELABUS FORMICARIUS: *Linn. Syſt. Nat.* 2. 620. 8.
Clerus formicarius. *Fab. Ent. Syſt.* 1. *p.* 207. 27. 5.
—— niger thorace rufo, elytris fafcia duplici alba bafique rubris. *Degeer. Inſ.* 5. 160. 3. *t.* 5. *f.* 8.

Only a few fpecies of this genus have been difcovered in this country; and neither of thofe are very common. We apprehend Attelabus Formicarius is rare, having only met with one fpecimen of it. It was found in May, on a fand-bank, near Coome Wood, Surry.

## F I G. III.　III.

### DERMESTES PELLIO.

*GENERIC CHARACTER.*

Antennæ terminated in a perfoliated club: the three extreme articulations thicker than the reft. Thorax convex: fcarcely margined. Head bent in ; and almoft concealed under the thorax.

*SPECIFIC*

# PLATE CCXXXI. 39

*SPECIFIC CHARACTER*

A N D

*SYNONYMS.*

Black. A white fpot on each wing-cafe.

Dermestes Pellio : niger elytris puncto albo. *Lin. Syft. Nat.*
2. 563.—*Fn. Sv.* 411.
Dermestes Pellio. *Fab. Ent. Syft.* 1. *p.* 228. 5.
*Oliv. Inf.* 2. 9. 11. 10. *tab.* 2. *fig.* 11.
*Schæff. Icon. tab.* 42. *fig.* 4.

———

A very common and deftructive creature. It infinuates itfelf into all kinds of fur, or the dried fkins of animals ; and in the ftate of larva, injures them confiderably. The larvæ of this tribe of Infects are numerous where they are fuffered to breed: they enter into and deftroy furniture, cloathing, and even food. Some fpecies are found upon the carcafes of animals ; while others, more tenacious of life, penetrate harder fubftances, and refift the camphor, verdigreafe, mufk, arfenic, and other drying or corroding fubftances, that prevent the depredations of moft Infects. One or two fpecies are in particular much to be dreaded by Collectors of Natural Curiofities : they per-forate the cabinet, or cafe, and when the larva is hatched, effect their deftruction. In collections of animals, birds, infects, and plants, they do great mifchief.

F 4 FIG.

## F I G IV. IV.

### DERMESTES SCARABÆOIDES.

*SPECIFIC CHARACTER*

AND

*SYNONYMS.*

Ovated. Black. Two red fpots on the wing-cafes.

DERMESTES SCARABÆOIDES. *Linn. Syft. Nat.* 2. 563. 17.—
  *Fn. Sv.* 428.
SFHÆRIDIUM SCARABÆOIDES: ovatum atrum elytris maculis
  duabus ferrugineis. *Fab. Ent. Syft.* 1. 77. 6. 1.

———————————

Found in dung. Fabricius feparates this fpecies from the Der-
meftides, and places it in a new genus *Sphæridium*.

P L A T E

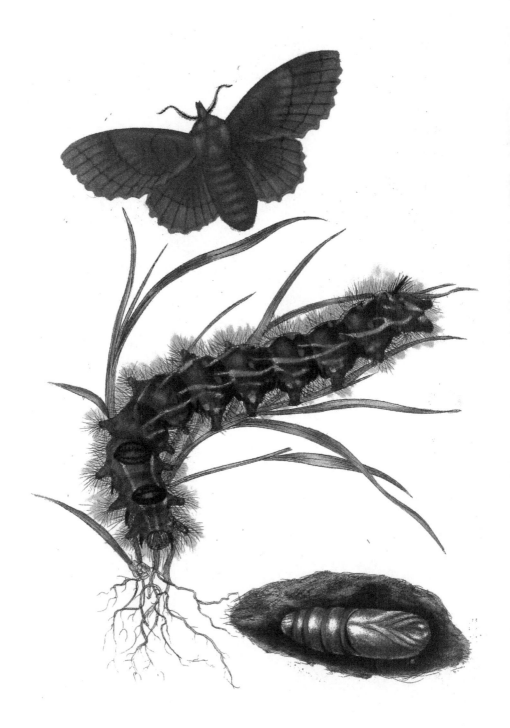

# 232

# PLATE CCXXXII.

## PHALÆNA QUERCIFOLIA.

### LAPPET MOTH.

### LEPIDOPTERA.

### GENERIC CHARACTER.

Antennæ taper from the bafe. Wings, in general, deflexed when at reft. Fly by night.

*Bombyx—Alis reverfis.*

### SPECIFIC CHARACTER

#### AND

### SYNONYMS.

Wings reverfed, fcalloped, red brown, with tranfverfe waved lines.

PHALÆNA QUERCIFOLIA: alis reverfis dentatis ferrugineis, ore
tibiifque nigris. *Linn. Syft. Nat.* 2. 812. 18.
*Fn. Sv.* 1110.

BOMBYX QUERCIFOLIA. *Fab. Ent. Syft. T.* 3. *p.* 1. 420. 42.
*Roef. Inf.* 1. *Phal.* 2. *tab.* 41.
*Schæff. Icon. tab.* 71. *fig.* 45.
*Merian Europ. tab.* 1. *fig.* 3.
*Reaum. Inf.* 2. *tab.* 23.
*Sulz. Inf. tab.* 16. *fig.* 93.
*Albin. Inf.* 1. *tab.* 16.

Phalæna

Phalæna Quercifolia is a rare and interesting Insect, and is the largest of the English bombyces, if we follow the arrangement of the *Entomologia Systematica* of Fabricius, and remove the Linnæan Bombyx Cossus\* to a new genus †.

The trivial name Lappet Moth has been given to this species by some early Collectors of English Insects, because they observed, when the creature was at rest, and the wings expanded in a natural position, the anterior part of the second pair *lapped* over the first, instead of the first pair laying on the second, as in many other species of Phalæna. This appearance is very striking, but not peculiar to Phalæna Quercifolia. *Bombyces*, with similar reversed wings, are numerous, and several of them natives of this country; as *Phal. Quercus*, and *Potatoria*.

In a former volume we have given The Pine Lappet Moth; an Insect that is extremely uncommon in Great-Britain; that, with the present species, are the only two British Phalæna called Lappet Moths; another Insect, closely allied to P. Quercifolia, and supposed to be the *Phal. Populifolia*, is said to be an English Insect, but on what authority, we are not informed. The very different appearances of the larva of Phal. Quercifolia in different stages of its growth, may possibly have caused some mistake; for in one skin they are brown, with whitish marks; in another greyish, with dark brown marks; and when of a full size, are more inclined to brown and grey in some specimens than in others. Notwithstanding, however, the variation of colours, in all its changes, we find that the two oval blue marks on the second and third segment of the body are constant, and sufficient to determine the species. The larva of Bombyx Populifolia is very similar to that of B. Quercifolia, when young, if we may judge by the only figure of it extant, but the mark across the second segment is narrow and black; that on the third segment broader, with two semi-lunated red spots.—*Vide Kleeman's Beytraege, &c. Vol.* 3. *tab.* 14.

---

\* Goat Moth.        † Cossus ligniperda.—*Fab.*

The

# PLATE CCXXXII.

43

The larva of P. Quercifolia we have taken in Darent Wood, Kent, on the grafs; it feeds alfo on Willow, Blackthorn, and Whitethorn. At the end of April, or not later than May, it forms a large and loofe fpinning interwoven with its hairs, of black, reddifh, and grey colours. The pupa is black, but appears perfectly white, being covered with a fine white pollen, or powder; each feg-ment is encircled by a belt, of a red colour. In July and Auguft it is found in the Fly ftate.

PLATE

# 233

# PLATE CCXXXIII.

## PHALÆNA PRUNARIA.

### PHOENIX MOTH.

### LEPIDOPTERA.

### GENERIC CHARACTER.

Antennæ taper from the base. Wings in general deflexed when at rest. Fly in the night.

*Geometra.*

### SPECIFIC CHARACTER

#### AND

### SYNONYMS.

Antennæ like a bristle. First pair of wings brown and grey, with two broad pale waved marks acrofs. Second pair, with waves on the posterior part.

PHALÆNA PRUNARIA: feticornis alis grifeo fufcis: fafciis duabus pallidis repandis: poftica femiterminali. *Linn. Syft. Nat.* 2. 869. 250.—*Fn. Sv.* 1267.—*Fab. Ent. Syft.* 3. *p.* 2. *p.* 178. 175.
*Wien. Verz.* 113. 19.
*Clerk. Phal. tab.* 7. *fig.* 3.
*Ammiral Inf. tab.* 23. *fig.* 1. 4.

----

Fabricius defcribes the larva of this rare Moth :—It is afh-coloured, with a black collar or mark on the neck : feet reddish brown, and the back fpotted with the fame colour.

The

The trivial Englifh name, Phœnix Moth, has been given to this Infect from a circumftance little known, and fcarcely deferving notice, except as it proves the impropriety of naming Infects from local circumftances, when any other can be well applied. A fmall part of a wood near London had been cut down, and a quantity of charcoal made on the fpot. This place had been often vifited by Aurelians, but the Phalæna Prunaria had never been difcovered there, nor indeed was then known as a Britifh Infect. On the following year, when the ground was cleared, and the underwood grown up, this Moth was found, it continued to be taken conftantly in the months of June and July for many years, in this place, and then totally difappeared. The late Mr. Bentley, known as a collector of Englifh Infects, difcovered a breeding-place of this Moth on Epping Foreft, and commonly found three or four fpecimens every feafon. We are not certain that it has been found in any other part of the kingdom. It feeds on the thorn, plumb and currant.

## F I G. II. II.

### PHALÆNA DUPLICATA.

### SPECIFIC CHARACTER

AND

### SYNONYMS.

Firft wings grey, with three tranfverfe waved lines.

PHALÆNA DUPLICATA: feticornis, alis grifeis, fafciis duabus trilineatis fufcis. *Fab. Ent. Syft.* 3. *p.* 2. *p.* 193. 234.

PHALÆNA PLAGIATA: feticornis, alis anticis canis: fafciis tribus trilineatis nigricantibus repandis. *Linn. Syft. Nat.—Fn. Sv. p.* 334. *n.* 1271.

Phal.

Phal. Plagiata. Das doppelte Band. *Berlin. Mag.* 4. *B. p.* 522.
    *n.* 38.
    *Schæff. Icon. tab.* 12. *fig.* 1. 2.
    *Clerk. Icon. tab.* 6. *fig.* 1.
    *Roßi Faun. Etr. T.* 2. *p.* 194. *n.* 1170.

Found in June. It is rare, and we believe has not been met with in the ftate of larva in this country. Foreign authors fay the larva is brown, variegated with red, and has a yellow line on each fide. *Kleemann* has not figured the larva with the Moth in his Supplement of the rare Infects found in Germany.

## FIG. III.

## PHALÆNA VESPERTARIA.

### SPECIFIC CHARACTER

AND

### SYNONYMS.

Antennæ feathered. Wings yellowifh : two dark waved ftreaks acrofs the firft pair ; one on the fecond pair : the fpace between the ftreaks and margins of the wings, dark.

PHALÆNA VESPERTARIA pectinicornis alis flavefcentibus : ftrigis
    duabus ; pofteriore limbum obfcurum difterminante.
    *Linn. Syft. Nat.* 2. 864. 224.
    *Fab. Ent. Syft.* 3. *p.* 2. 149. 74.
Phalæna parallelaria. *Wien. Verz.* 104. 15.

Found in Hornfey-Wood in July, and alfo in Norwood.

FIG.

# FIG. IV.

## PHALÆNA CHÆROPHYLLATA.

### GREAT CHIMNEY-SWEEPER.

*SPECIFIC CHARACTER*

AND

*SYNONYMS.*

Antennæ like a briftle: Wings black erect: firft pair white at the tips.

PHALÆNA CHÆROPHYLLATA feticornis atra alis erectis : anticis apice albis. *Linn. Syft. Nat.* 866. 237.—*Fab. Ent. Syft. I.* 3. *p.* 2. 184. 200. *Wien. Verz.* 116. 1.

———

Appears in the Winged ftate late in July. Is produced from a reen capillar, which feeds on *Cherophyllum Silveftre,* or wild cicely.

PLATE

# 234

# PLATE CCXXXIV.

## TENTHREDO LUTEA.

### YELLOW SAW-FLY.

#### HYMENOPTERA.

Wings four, generally membraneous. Tail of the females armed with a fting.

### GENERIC CHARACTER.

Without probofcis. Mouth armed with jaws. Sting compofed of two laminæ, dentated, like a faw, and almoft concealed within the abdomen. Two tubercles on the fcutellum.

### SPECIFIC CHARACTER

AND

### SYNONYMS.

Antennæ clubbed, yellow. Abdomen yellow, except the fecond fegment, which is black.

TENTHREDO LUTEA : antennis clavatis luteis, abdominis fegmentis
plerifque flavis. *Linn. Syft. Nat.* 2. 921. 3.—*Fn. Sv.* 1534.
*Fab. Ent. Syft. I.* 2. *p.* 105. 138. 3.
*Roef. Inf.* 2. *Vefp. tab.* 13.
*Schæff. Icon. tab.* 103. *fig.* 2. 3.
*Degeer Inf.* 2. 2. 223. 7. *tab.* 33. *fig.* 8. 16.

G                                        Very

Very uncommon in this country. The larva has been found on the Willow, but unlefs taken when ready to become a pupa, it is impoffible to rear it to the winged ftate. We are little acquainted with the peculiar habits of thefe Infects, and cannot therefore feed them in a proper manner.

Moft of the Tenthredines enfhroud themfelves in a net-work covering, and remain in the earth till the Fly burft forth ; others faften the web againft the branches of trees, or on the trunk near the earth. The habits of Tenthredo lutea are very fimilar to thofe of T. Vitellinæ ; the winged Infect appears in June.

P L A T E

# PLATE CCXXXV.

## CHRYSIS CYANEA.

### HYMENOPTERA.

Wings four: generally membraneous. Tail of the females armed with a fting.

### GENERIC CHARACTER.

No probofcis. Armed with jaws. Antennæ filiform.

### SPECIFIC CHARACTER

#### AND

### SYNONYMS.

Very gloffy blue green. End of the abdomen furnifhed with three teeth.

CHRYSIS CYANEA: glabra nitens thorace abdominifque cœruleis, ano tridentato. *Linn. Syft. Nat.* 2. 948. 5.—*Fn. Sv.* 1667.
*Fab. Ent. Syft.* 2. 147. *p.* 243. 20.
Vefpa cœrulea nitens. *Geoff. Inf.* 2. 484. 23.—*Schæff. Icon.* tab. 81. *fig.* 5.

---

The natural fize of this Infect is given at Fig. I. in the annexed plate. It is far inferior in beauty to either Chryfis ignita or bidentata, figured in the early part of this work; but as the genus is very limited, we have a given figure of this fpecies. It is very abundant on all k nds of fruit-trees in the fummer.

PLATE

# PLATE CCXXXVI.

## FIG. I. I.

### PAPILIO CORYDON.

#### CHALK-HILL BLUE BUTTERFLY.

*GENERIC CHARACTER.*

Antennæ clubbed. Wings erect when at reft. Fly by day.

*SPECIFIC CHARACTER*

AND

*SYNONYMS.*

Wings entire, above filvery or pale blue, with a black margin. Beneath grey, with eye-fhaped fpots.

HESPERIA CORYDON: alis integris cœruleo argenteis: margine nigro, fubtus cinereis: punctis ocellaribus, pofticis macula centrali alba. *Fab. Ent. Syft.* 3. *p.* 1. *p.* 298. 133.

PAPILIO CORYDON. *Wien. Verz.* 184. 10.
PAPILIO CORYDON. *Efp. pap. tab.* 33. *fig.* 4.
PAPILIO TIPHYS. *Efp. pap. tab.* 51. *fig.* 4.

Found on the chalk-hills between Dartford and Rochefter; particularly on a long range of hillocks leading from Dartford to the wood of Darent. Hence the Butterfly has been called the Chalk-hill blue. We believe it has not been found in any other part near London. The larva is unknown, it appears in the winged ftate, the firft and fecond week in July.

H                                                    FIG.

## F I G. II. II.

### PAPILIO LINEA.

#### SMALL SKIPPER BUTTERFLY.

*SPECIFIC CHARACTER*

AND

*SYNONYMS.*

Wings entire, brown, divaricated, margin black.—An oblique black mark on the anterior wings.

HESPERIA LINEA: alis integerrimis divaricatis fulvis: margine nigro. *Fab. Ent. Syft.* 3. *p.* 1. 326.
PAPILIO LINEA. *Wien. Verz.* 159. 5.
PAPILIO THAUMAS. *Efp. pap. tab.* 36. *fig.* 2. 3.
PAPILIO SYLVESTRIS. *Pod. Muf.*

––––––––––––––––

A very generally diffufed fpecies, but not common ; it is fimilar to the Papilio Sylvanus of Linnæus, or Hefperia Sylvanus of Fabricius, which is found in the greateft abundance in the fkirts of woods in fummer. Its metamorphofe is unknown.

PLATE

# 237

# PLATE CCXXXVII.

## PHALÆNA BETULARIA.

### PEPPERED MOTH.

## GENERIC CHARACTER.

Antennæ taper from the bafe. Wings in general deflexed when at reft. Fly by night.

## SPECIFIC CHARACTER

### AND

## SYNONYMS.

Antennæ feathered. Wings entirely white, fpeckled with black; a black bar acrofs the thorax.

PHALÆNA BETULARIA : pectinicornis, alis omnibus albis, thorace fafcia nigra, antennis apice fetaceis. *Linn. Syft. Nat.* 2. 862. 217. *Fn. Sv.* 1287. *Fab. Spec. Inf.* 2. 252. 56.

PHALÆNA antennis pectinatis, alis horizontalibus albis nigro punctatis maculatifque, thorace fafciato. *Degeer. Inf. Verz. Germ.* 2. 1. 250. 1 *tab.* 5. *fig.* 18. *Ammiral. Inf. tab.* 21. *Schæff. Icon. tab.* 88. *fig.* 4. 5. *Albin. Inf. tab.* 91, 92. *Kleman. Inf.* 1. *tab.* 39. *fig.* 6.

H 2                                                          Found

Found on the Lime, Willow, and Elm in the ſtate of Larva, changes to the Pupa in September ; and the Moth appears in May. The Larva of this creature differ very much in their ſhades of colour ; they are generally blackiſh or dark olive with a few obſcure red ſpots.

PLATE

# PLATE CCXXXVIII.

## FIG. I. I.

### PAPILIO HYALE.

CLOUDED YELLOW BUTTERFLY.

LEPIDOPTERA.

### GENERIC CHARACTER.

Antennæ clubbed.  Wings erect when at reft.  Fly by day.

### SPECIFIC CHARACTER

AND

### SYNONYMS.

Wings rounded, yellow: an orange fpot on the pofterior wings: beneath, a large filver fpot, with a fmall contiguous fpot of the fame.

PAPILIO HYALE: alis rotundatis flavis: pofticis macula fulva; fubtus puncto fefquialtero argentes. *Linn. Syft. Nat.* 2. 764. 100.

―――――――――

Though we cannot but admire the Linnæan definitions, for their perfpicuity in general; we muft in fome inftances blame him for that inattention which has betrayed fucceeding naturalifts into errors,

H 3 and

and even abfurdities. The force of this remark, however harſh it may appear, will apply in a particular degree to the ſpecific definition and quoted Synonyms of Papilio Hyale, and conſequently to two other ſimilar ſpecies involved in the ſame error.

Linnæus gave the deſcription of Papilio Hyale, as above quoted in the Syſtema Natura, from an infect in his own cabinet, and quotes, in the Synonyms, the Butterfly figured by *Roeſel, Vol.* 3. *tab.* 46. *fig.* 4. 5. The works of that author being known in every part of Europe, the entomologiſts of that time received the figure as that of the true Hyale; and relying on the accuracy of the Linnæan references, the miſtake has been overlooked to the preſent period. Fabricius, who is the lateſt ſyſtematic writer on this ſcience, quotes the figures in Roeſel, as Linnæus had himſelf in the firſt inſtance; and he alſo refers to figures of the ſame infect in the works of Cramer and Schæffer. An error of ſuch ſpecious appearance could only be detected by a reference to the ſpecimen in the Linnæan Cabinet, at this time in the poſſeſſion of Dr. Smith, and by this it appears that every author has miſtaken the ſpecies of Linnæus, and that Linnæus was himſelf miſtaken in ſuppoſing the infect deſcribed was the ſame as that figured by Roeſel, and to which he refers: that the Linnæan *Papilio Hyale* is what later authors have conſidered *Papilio Palæno*, and that the true *Papilio Palæno* is not a Britiſh ſpecies.

Theſe errors are ſo complicated that we muſt examine the characters aſſigned to each ſpecies with the utmoſt attention, and we ſhall then find his deſcriptions correct, but the ſynonyms erroneous. Papilio Hyale is deſcribed with *yellow wings*; the colour of the wings in the ſuppoſed Hyale is not of that kind which Linnæus would have called yellow, *(flavis,)* but *fulvis*, being of a deep orange colour, much inclining to red. Either of theſe expreſſions will certainly admit of great latitude, but we muſt not therefore confound one with the other. The two ſilver ſpots are not conſtant, though it forms a part of the ſpecific character; we have ſeen in both the clouded yellow and clouded orange butterflies, ſometimes only one ſpot, though in general it has two. On the whole, the Linnæan

deſcrip-

PLATE CCXXXIX.    59

defcription of *Papilio Hyale* feems to agree with the fictitious
*P. Palæno* of our collections, and the fpecimen in the Linnæan
cabinet places it beyond conjecture.

Fabricius has not attended to the errors of former authors on this
fubject ; even in his laft work, *Syft. Ent.* he adds to the fpecific defcription of his P. Hyale, Mas margine alarum nigro immaculato, fœmina maculato\*, by this it is evident he alludes to the
clouded orange, for it is not fo in the Linnæan infect ; the broad
bar of black being conftantly fpotted in both fexes.   We have alfo
obferved that the rare variety with white wings is only the female ;
that which is yellow is the male : the fame is obferved alfo of
Papilio Rhamni, or Brimftone Butterfly ; and as the males of all
infects are more abundant than the females, and the males of P. Hyale
are rare, the variety, or fex with white wings muft be extremely fo.
Thefe have been taken in a clover field in the month of Auguft, in
company with the clouded orange.

It may be proper to clofe this defcription with a few obfervations
on the true Papilio Palæno, as the fubject before us has hitherto
paffed under that name.   Linnæus fays, alis integerrimis flavis apice
nigris margineque fulvis : pofticis fubtus puncto argenteo ; this does
not agree with, or at leaft exprefs the Infect generally called Palæno,
and the fpecimen in the Linnæan cabinet proves it to be a different
fpecies ; the P. Palæno has no yellow fpots on the black margins of
the wings, and the fpot in the center of the anterior pair is fmall
and fhaped like a fhuttle ; its native place is unknown.   A variety
of the pale clouded yellow is found in America, and by miftake is
fometimes placed in cabinets as a Britifh fpecies.

---

\* The wings in the males have a broad black bar on the exterior margin ; in the female the
bars are fpotted.

H 4

FIG.

# F I G. II.

## PAPILIO EDUSA.

### CLOUDED ORANGE BUTTERFLY.

#### LEPIDOPTERA.

#### PAPILIO.

## *SPECIFIC CHARACTER*

### AND

## *SYNONYMS.*

Wings entire, fulvous, or orange, with a black fpot and margin of the fame colour. On the under fide greenifh; a black fpot on the anterior wings; filver on the pofterior wings.

PAPILIO EDUSA: alis integerrimis fulvis: puncto margineque nigris, fubtus virefcentibus: anticis puncto nigro, pofticis argenteo. *Fab. Ent. Syft.* 3. *p.* 2. 206. 643.

——————————

If our obfervations on the preceding fpecies are fatisfactory and conclufive, the Infect before us muft be a diftinct fpecies, and not the true P. Hyale, for which it has ever been received. This involves another interefting confideration, for Linnæus muft have been acquainted with this fpecies, as he refers to figures of it in feveral Authors. It is probable he confidered it as a variety, for it does not appear under another name in any part of his Writings.

9

Fabricius

# PLATE CCXXXVIII. 61

Fabricus has defcribed a new fpecies of butterfly, under the name *Edufa*, in his laft work *Ent. Syft.* which feems to agree with our infect, and we are confirmed in our opinion by Mr. Jones of Chelfea, who affifted Fabricius with confiderable information, and affures us it is certainly the P. Edufa of that author.

In a former part of this work we have given the male of this infect as P. Hyale: the annexed figure reprefents the female, having large yellow fpots in the black border of the wings. Both this and the preceding fpecies are figured by Efper, in the *Papiliones de l'Europe*; and by *Schæffer*, in the *Icones Rabifbon*, &c.

PLATE

# PLATE CCXXXIX.

## FIG. I.

### PHALÆNA TREPIDA?

SWALLOW PROMINENT MOTH.

LEPIDOPTERA.

#### GENERIC CHARACTER.

Antennæ taper from the bafe. Wings in general deflexed when at reft. Fly by night.

#### SPECIFIC CHARACTER

AND

#### SYNONYMS.

Wings deflexed ; a prominence on the back. Anterior wings pale in the middle ; brown next the margin ; ftreaked. A fpot in the center of the wing.

BOMBYX TREPIDA : alis deflexis dorfo unidentatis : puncto medio ocellari ftrigaque poftica maculari fufcis. *Fab. Ent. Syft.* 3. *p.* 1. 449. 130.

Bombyx tremula. *Wien. Verz.* 49. 4.

The Swallow Prominent Moth is fcarce, the larva is fuppofed to live under the bark of willows, but it is more certain that the Moth is feldom found, except among thofe trees. In the day-time it has been feen againft the trunk of trees, in the manner reprefented in the Plate.

We quote the authority of Fabricius with diffidence ; his character is ambiguous ; and can only be defined by the very general defcription he has added to it.

FIG.

## FIG. II.

### PHALÆNA COMPRESSA.

LEPIDOPTERA.

BOMBYX.

*SPECIFIC CHARACTER*

AND

*SYNONYMS.*

Wings compreffed; white, with a large brown mark continued acrofs the anterior Wings; grey in the middle, with feveral lunar white marks.

BOMBYX COMPRESSA : alis compreffo adfcendentibus niveis : ma-
cula communi fufca, centrali grifea : lunula alba.
*Fab. Ent. Syft.* 3. *p.* 2. 455. 149.
Phalæna fpinula. *Wien. Verz.* 64. 6.
*Panz. Faun. Germ.* 1. *tab.* 6.

Not very uncommon in the month of June; it is called the Goofe-egg Moth.

PLATE

# PLATE CCXL.

## MELOE TECTA.

### COLEOPTERA.

### GENERIC CHARACTER.

Antennæ moniliform, extreme articulation oblong. Thorax roundifh. Elytra foft and flexible. Head inflected and gibbous.

### SPECIFIC CHARACTER

#### AND

### SYNONYMS.

Black. Wing-cafes nearly the length of the Abdomen. Antennæ thickeft in the middle.

MELOE TECTA : atra, elytris abdomine haud brevioribus, antennis medio craffiffimis. *Panz. Faun. Inf. Germ.*
Der Maywurmkäfer mit ungewöhalich langen Flügeldecken.— *Panz. Ibid.*

———————

This is a rare Infect, and has not been defcribed by Linnæus or Fabricius. It bears a diftant refemblance to *Meloe Profcarabæus*; but the fingular ftructure of the Antennæ will alone prove it a diftinct fpecies. The Antennæ in Meloe Profcarabæus are moniliform, or compofed of feveral bead-like articulations, nearly of the fame fize : thofe of this new fpecies are thickeft in the middle ; and in one point of view the fourth, fifth, and fixth articulations appear remarkably large and globular : in another, they feem writhed or diftorted, and very concave ; the other joints are as in the former fpecies.—The whole Infect has a fhining blue glofs : Meloe Pro-

fcarabæus

fcarabæus is coal black. It is fmaller, and the Wing-cafes nearly cover the Abdomen. In Meloe Profcarabæus the wing-cafes are only one-third the length of the Abdomen.

In fome Cabinets, this Infect is arranged with a new fpecific name *autumnalis* ; but as we find it is not a non-defcript, we prefer that, under which it has been already defcribed. In this we not only avoid the confufion arifing from a change of names, but reject one merely local for another expreffive of its fpecific character.

Our Specimens were found on Epping Foreft, in July.

PLATE

# PLATE CCXLI.

## SPHINX POPULI.

### POPLAR HAWK MOTH.

#### LEPIDOPTERA.

### GENERIC CHARACTER.

Antennæ thickeft in the middle. Wings, when at reft deflexed. Fly flow, morning and evening only.

### SPECIFIC CHARACTER

#### AND

### SYNONYMS.

Wings reverfed, dentated, grey: a white central fpot on the anterior Wings. Pofterior Wings red at the bafe.

SPHINX POPULI: Alis dentatis reverfis grifeis: anticis punɛto albo, pofticis bafi ferrugineis. *Linn. Syft. Nat.* 2. 797. 2.—*Fn. Sv.* 1084.
*Roef. Inf.* 3. tab. 30.
*Schæff. Icon. tab.* 100.
*Degeer. Inf.* 1. tab. 8. *fig.* 5.
*Sepp. Inf.* 3. 3. tab. 1.
*Albin. Inf. tab.* 38. *fig.* C.
*Wilks pap.* 11. tab. B. C.

---

This beautiful Infeɛt is very common in this country, and not lefs fo in every other part of Europe: it feeds on the poplar and
willow

willow in the larva ftate, and frequents thofe trees in the winged
ftate alfo. About the month of September, the Larva or Caterpillars are full grown, and change to the Pupa : the Sphinx appears
in May.

PLATE

# PLATE CCXLII.

## PAPILIO CINXIA.

PLANTAIN FRITILLARY.

LEPIDOPTERA.

### GENERIC CHARACTER.

Antennæ clubbed at the end. Wings erect, when at rest. Fly by day.

### SPECIFIC CHARACTER

AND

### SYNONYMS.

Wings dentated, brown, with black marks: beneath fulvous, with three whitish bands acrofs the lower Wings, marked with black fpots.

PAPILIO CINXIA: Alis dentatis fulvis nigro maculatis: pofticis fubtus fafciis tribus albidis nigro maculatis. *Linn. Syft.* 2. 784. 205.—*Sv.* 1063. *Fab. Ent. Syft.* 3. *p.* 2. 250. 779. *Roef. Inf.* 4. *tab.* 13. *fig.* 4. 5. *Geoff. Inf.* 2. 45. 12. *Wilks pap.* 58. *tab.* 3. *a* 8. *Efp. pap.* 1. *tab.* 16. *fig.* 2. *Schæff. Icon. tab.* 204. *fig.* 1, 2.

β. Papilio Delia, alis dentatis fulvo nigroque variis: pofticis fupra punctis quatuor ocellaribus, fubtus albis: fafciis duabus fulvis; pofteriore nigro punctata. *Linn. Fab.*

Papilio Delia. *Wien. Verz.* 179. 6.

I

The

The Larva are black, befet with fpines and tufts of the fame colour: the fides are marked with a double row of white fpots, the feet red. It is found on the long plantain in April. The Flies appear in May. This is the rareft of the Britifh Fritillary Butterflies, if we except Papilio Lathonia, the Queen of Spain Butterfly.

## F I G.  II.

### PAPILIO LUCINA.

#### DUKE OF BURGUNDY FRITILLARY.

*SPECIFIC CHARACTER*

AND

*SYNONYMS.*

Wings indented, dark brown with bright yellowifh-brown fpots. Two rows of white fpots on the underfide of the pofterior wings.

PAPILIO LUCINA: Alis dentatis fufcis teftaceo maculatis: fubtus
fafciis duabus macularum albidarum. *Linn. Syft.*
*Nat.* 2. 784. 203.—*Fn. Sv.* 1001.
*Fab. Ent. Syft.* 3. *p.* 1. 250. 778.
*Raj. Inf.* 122. 12.
*Schæff. Icon. tab.* 172. *fig.* 1. 2.
*Petiv. Gazoph. tab.* 16. *fig.* 10.

This pretty Infect is found in the winged ftate in May; the Larva is unknown.—Taken in Hornfey wood.

PLATE

# PLATE CCXLIII.

## FIG. I. I.

### COCCINELLA 14 GUTTATA.

*GENERIC CHARACTER.*

Antennæ fubclavated, truncated. Palpi club-formed, extreme articulation heart-fhaped. Body hemifpherical. Thorax and elytra margined.

*SPECIFIC CHARACTER*

AND

*SYNONYMS.*

Red, with fourteen white fpots.

COCCINELLA 14 GUTTATA: coleoptris rubris, punctis albis qua-tuordecim. *Linn. Faun. Suec.* 492. *Syft. Nat. p.* 583. *n.* 34. *Fab. Ent. Syft.* 1. *p.* 284. *n.* 85.

---

This fpecies is rather of a brown than red colour, as defcribed by Linnæus. It is probably an uncommon Infect. The fmalleft figure 1 is the natural fize.

K                    FIG.

# PLATE CCXLIII.

## FIG. II. II.

### COCCINELLA ANNULATA.

*SPECIFIC CHARACTER*

AND

*SYNONYMS.*

Red. An oblong black ring acrofs the wing-cafes.

COCCINELLA ANNULATA: coleoptris rubris, macula fubannulari nigra. *Linn. Syft. Nat.* p. 579. *n.* 5. *Fab. Ent. Syft.* 1. *p.* 268. *n.* 14.

―――――

Not a very common fpecies.

―――――――――――――――

## FIG. III. III.

### COCCINELLA CASSIDOIDES,

*SPECIFIC CHARACTER.*

Black with a red lunular mark, and a round fpot of the fame colour, on each of the wing-cafes. Margin prominent.

COCCINELLA CASSOIDES: elytris nigris, lunulâ punctoque rubris, margine prominulo. *Marfham MS.*

―――――

This is a nondefcript Infect. It was found in May.

PLATE

# PLATE CCXLIV.

## PAPILIO CAMILLA.

### WHITE ADMIRABLE.

#### GENERIC CHARACTER.

Antennæ clubbed at the end. Wings erect, when at reft. Fly by day.

#### SPECIFIC CHARACTER

#### AND

#### SYNONYMS.

Wings dentated. Above, uniform dark brown, with a white band, and fpots. Inner angle of the pofterior wings red.

PAPILIO CAMILLA: alis dentatis fufcis fubconcoloribus albo fafciatis maculatifque, angulo ani rubro. *Linn. Syft. Nat.* 2. 781. 187. *Roef.* 3. *tab.* 33. *fig.* 3. 4.

———

The White Admirable Butterfly feeds upon the common honey fuckle or woodbine, and is found in the winged ftate in the months of June and July, in the fkirts of woods; its habit is much the fame as that of P. Atalanta, Red Admirable, but it is by no means fo common.

K 2                                                    This

This species has hitherto been deemed the Papilio Camilla of Linnæus, though it differs in a flight degree from the descriptions and figures of authors who describe only German or Swedish specimens of it. In the late editions of the *Systema Naturæ*, P. Camilla is described with P. Sibilla, a Papilio nearly allied to it, but which Linnæus considered as a distinct species; his description of *Camilla* expressly saying " angulo ani rubro."—The angulis ani, of P. Sibilla *, not being of a red colour, removes it from the English species.

Fabricius is of a different opinion, and in dissenting from his authority it is incumbent to state our objections. In the *Species insectorum* of that author, the P. Camilla with the Linnæan specific character, is made a variety β of Sibilla. The synonyms of the two insects are ambiguous, and the references not more satisfactory. Among others he refers for P. Sibilla, *to Drury's Inf. 2. tab.* 16. *fig.* 1, 2,—*to Roefel Inf.* 3. *tab.* 70. *fig.* 1, 2, 3,—*and to Schæffer* 152 †. *fig.* 1, 2. The two last are perhaps the same species: the first is unquestionably different. For *P. Camilla* he refers to *Roefel tab.* 33. *fig.* 3, and this agrees with our specimen, except in the colours of the upper surface being somewhat paler; so that we may conclude our English Insect is not only the P. Camilla of Linnæus but also the β Camilla of Fabricius.

The ultimate opinion of Fabricius is however different, for in the *Entomologia Systematica* since published, Camilla and Sibilla stands a distinct species; and Camilla is thus described, " alis dentatis atris coeruleo micantibus: fascia utrinque maculari alba, posticis subtus basi argentea immaculata." *Linn. Syst. Nat.* 2. 781. 187. *habitat in Austria.* Thus P. Camilla will no longer agree with our insect, nor with that figured by *Roefel, plate* 33. *fig* 3. But if we refer to the Systema Naturæ of Linnæus, we discover another error, for

---

* Roefel's figure, vol. 3. tab. 70. has an obscure red band entirely across the posterior wings.—Schæffer's figure has no trace of red on that part.

† Erratum.—Designed for 153—for 152 is the German variety of Papilio Iris, Purple Emperor Butterfly.

instead

# PLATE CCXLIV.    77

inftead of the above quoted fpecific character, Linnæus only fays, " Alis dentatis fufcis fubconcoloribus albo fafciatis maculatifque, angulo ani rubro." *p.* 781. 187. and this is clearly our infect.— From this the whole of the Fabrician account is obvioufly a complicated error of defcription and fynonyms, and without attempting to inveftigate it further, we fhall fpeak of both fpecies as they appear to us.

We confider the Common Englifh Admirable, as the true *P. Camilla*; and that Sibilla, and not Camilla is the Auftrian fpecies, as we have received it from that country; it is *much darker* in the *upper furface*, and has a *row* of *fhining blue fpots* all round the wings: the *bafe* of the *pofterior wings*, beneath *filvery* and *without fpots*, as Fabricius defcribes his Camilla. Our infect on the contrary has *no row of blue fpots* on the *upper furface*, but a *red fpot* at the inner angle of the pofterior wings; the *bafe* of the *lower wings* are alfo *filvery beneath*, but *has black fpots* upon it. They differ in many other refpects: thefe alone determine them to be two fpecies, and ours to be the *Camilla* of Linnæus.

We have infpected the drawings of Mr. Jones, from which Fabricius defcribes moft of his Papiliones, and are confirmed in our opinion.

K 3                                    PLATE

245

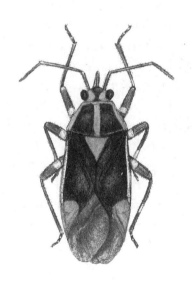

# PLATE CCXLV.

## CIMEX FLAVO-MARGINATUS.

### *GENERIC CHARACTER.*

Roftrum inflected. Antennæ longer than the thorax, wings folded crofswife, thorax margined. Feet formed for running.

### *SPECIFIC CHARACTER.*

Black, longitudinal line on the thorax. Scutellum, exterior margin of the elytra, and fpot in the apex yellow.

CIMEX FLAVO-MARGINATUS niger, thoracis lateribus lineaque dorfali, fcutello elytrorumque margine apiceque macula flavis.

———

A nondefcript fpecies: it was found on a thiftle.

The line at fig. 1 denotes the natural fize.

K 4                              PLATE

# PLATE CCXLVI.

## FIG. I.

### PHALÆNA PARTHENIAS.

LEPIDOPTERA.

### GENERIC CHARACTER.

Antennæ taper from the base. Wings in general deflexed when at rest. Fly by night.

### SPECIFIC CHARACTER

AND

### SYNONYMS.

Wings deflexed. First pair with greyish waved bars acrofs. Second pair orange-red, having a large triangular black spot at the interior edge.

PHALÆNA PARTHENIAS: *Linn. Syst. Nat.* 2. 835.94. *Fn. Sv.* 1160.

BOMBYX VIDUA, alis deflexis fufcis: anticis cinereo fubfafciatis, pofticis macula bafeos fafciaque rufis. *Fab. Ent. Syst. T.* 3. *p.* 1. 468, 190.

---

This fpecies was defcribed by Linnæus as *Phalæna Parthenias*; Fabricius alters it to *Vidua*. It is not figured by any author unlefs the *Ph. Geom.* Glauco fafciata Gözii, figured in *Kleman. T.* 1. *tab.* 40. *fig.* 4, 5, be the fame infect.

Found

Found on the white Poplar. (Populi Alba). It is a rare Infect and has been taken in Hornfey Wood in May.

---

# F I G. II.

## PHALÆNA DUBITATA.

### TISSUE MOTH.

### *SPECIFIC CHARACTER.*

Antennæ fetaceous, wings waved with brown grey and black. The nerves of the wings fpotted with white.

PHALÆNA DUBITATA, feticornis alis obfcuris fufco cinereoque
undatis: nervis albo punctatis. *Fab. Ent. Syft. T.*
3. *p.* 2. 168. 141.
Phalæna dubitata. *Wien. Verz.* 109. 3.

---

The Tiffue Moth is rarely met with near London. Our fpecimen was taken near Bath.

P L A T E

247

# PLATE CCXLVII.

## PAPILIO PAPHIA.

### SILVER STRIPE FRITILLARY BUTTERFLY.

### LEPIDOPTERA.

### *GENERIC CHARACTER.*

Antennæ clubbed at the ends. Wings erect, when at reft. Fly by day.

### *SPECIFIC CHARACTER*

### AND

### *SYNONYMS.*

Wings dentated fulvous, fpotted with black, under-fide ftriped with filver.

PAPILIO PAPHIA : Alis dentatis fulvis nigro maculatis, fubtus faf‑
ciis argenteis. *Linn. Syft. Nat.* 2. 786, 209.
*Fn. Sv.* 1064.
*Geoffr. Inf.* 2. 42. 8.
*Roef. Inf.* 1 *pap.* 1 *tab.* 7.
*Schæff. Icon. tab.* 27. *fig.* 3. 4.
*Wilks pap.* 57. *tab.* 2, *a.* 7.

---

Papilio Paphia is an elegant fpecies of the Britifh Papiliones. In fize, colour, and general appearance of the upper fide it is very fimilar to Papilio Agala; in the underfide, it is extremely different.
Both

Both of thefe Butterflies are remarkable for that pecular fhining appearance of polifhed filver with which a few other of the european
Fritillary Butterflies are ornamented; but in Papilio Agala, this
filver is difpofed in diftinct fplafhes or fpots, while in Papilio Paphia
it appears in tranfverfe ftreaks. Thefe ftreaks are finely foftened
into the red and olive green of the wings, and produce altogether a
fingular and charming effect. It is from the latter circumftance
the early Englifh collectors termed this the *filver-wafh* Fritillary.

The Caterpillar of this butterfly is found on the grafs in May. It
is of a plain yellowifh brown, with feveral longitudinal ftripes of
dark brown; it is alfo thickly befet with barbed fpines, a quarter of
an inch in length, and has in particular two of a remarkable form on
the firft annulation next the head. It remains in the chryfalis twenty
or twenty one days; and appears in the winged ftate early in June.

P L A T E

# PLATE CCXLVIII.

## FIG. I. II.

## PHALÆNA ATOMARIA.

### Dark Heath Moth.

### *GENERIC CHARACTER.*

Antennæ taper from the bafe. Wings in general deflexed when at reft. Fly by night.

### *SPECIFIC CHARACTER*

### AND

### *S Y N O N Y M S.*

Antennæ feathered. Wings entirely yellowifh ; with bands of brown, and the whole thickly fprinkled with atoms of the fame colour.

PHALÆNA ATOMARIA : pectinicornis ; alis omnibus lutefcentibus
      fafciis atomifque fufcis. *Linn. Faun. Suec.* 1245.—
      *Syft. Nat. p.* 862. *n.* 214.
Phalæna ifofelata. *Scop. carn.* 558. *female.*
Phalæna pennata. *Scop. carn.* 569. *male.*

---

The Dark Heath Moth is confidered as the Phalæna Atomaria of Linnæus, in the *Faun. Suec.* and *Syftema Naturæ.* Fabricius adopts the Linnæan character, though in the Synonyms it is confounded with another *Geometræ,* the *Hirtaria* of fome entomologifts. He refers to the *Phalæna iffofelata* and *Phalæna pennata* of Scopoli, which are evidently the two fexes of the Dark Heath Moth; but the Infect figured in Plate XXXIV. *Kleman's Inf.* is very diftinct, and is the P. Hirtaria before alluded to. We cannot but obferve how clofely the Linnæan defcription will apply in this inftance, to two Infects which at firft fight appear fo extremely different.

This fpecies is very common on heaths in May. The male is much darker in colour than the female, and has curious pectinated and ciliated antennæ, as Scopoli obferves. The female has been miftaken for a diftinct fpecies.

---

## F I G.  III.

### PHALÆNA CLATHRATA.

#### PALE HEATH MOTH.

*SPECIFIC CHARACTER*

AND

*SYNONYMS.*

Antennæ fetaceous. Wings entirely yellowifh, with tranfverfe and longitudinal dark lines interfecting each other.

PHALÆNA CLATHRATA: feticornis alis omnibus flavefcentibus: lineis nigris decuffatis. *Lyn. Syft. Nat.* 2. 867. 238. —*Fn. Sv.* 1275.—*Fab. Ent. Syft.* 3. *p.* 2. *p.* 183. 194.
*Schæff. Icon. tab.* 216. *fig.* 2. 3.
*Scop. Carn.* 536.
*Sulz. Hift. Inf. tab.* 23. *fig.* 2.

---

This is a rare infect, except in Kent, where Dr. Latham informs us it is more common than the preceding fpecies. Both fexes are very fimilar to the female Dark Heath Moth. The larva is unknown, but is fuppofed to feed on the *Erica*, being always found on thefe plants in the winged ftate. Thefe infects have been called Heath Moths becaufe they live in heaths, chalk-pits, and other barren places.—The Moths appear in June.

PLATE

# PLATE CCXLIX.

## FIG. I. I.

### CURCULIO RUFUS.

*GENERIC CHARACTER.*

Antennæ fubclavated, feated in the fnout or probofcis, which is prominent.

*SPECIFIC CHARACTER.*

Snout long. Feet formed for leaping, entirely red, except the eyes, which are black.

CURCULIO RUFUS: longiroftris pedibus faltatoriis, totus rufis, oculis nigris folis exceptis. *Marfh. MS.*
*Geoff. Inf.* 1. 286. 19. 2.

---

Probably a new fpecies; it differs from the *Curculio Quercus* of the *Faun Suec*, and *Curculio viminalis* of *Ent. Syft.* in having the abdomen red, and being rather larger.

---

## FIG. II. II.

### CURCULIO ALNI.

COLEOPTERA.

*SPECIFIC CHARACTER.*

Snout long. Feet formed for leaping. Elytra livid red, with two obfcure fpots.

CIRCULIO

Curculio Alni: longiroftris, pedibus faltatoriis elytris lividis, maculis duabus obfcuris. *Faun. Suec* 608.—*Fab. Ent. Syft. I.* 445. 216.

Feeds on the leaves of the Alder.

The fmalleft infect at Fig. 2. reprefents the natural fize.

## FIG. III.

### CURCULIO TENUIROSTRIS.

*SPECIFIC CHARACTER.*

Snout long, thighs dentated, black, a fhort tranfverfe white ftripe on the elytra. Antennæ red.

Curculio tenuiroftris: longiroftris, femoribus dentatis niger, elytris albo fubfafciatis, antennis rufis. *Fab. Ent. Syft. I. p.* 2. 443. 204.

Defcribed by Fabricius from the cabinet of Sir Jofeph Banks, Bart. without a reference to any figure. His general defcription is, Minor. C. Ceraforum. Caput nigrum roftro tenui, atro, glabro. Antennæ rufæ clava cinerea. Thorax niger pilis breviffimis cinereis. Scutellum cinereum. Elytra nigra fafciis plurimis, undatis pallidioribus.

PLATE

# PLATE CCL.

## PAPILIO BETULÆ.

### Brown Hair Streak Butterfly.

### Lepidoptera.

### *GENERIC CHARACTER.*

Antennæ clubbed. Wings erect when at reft. Fly by day.

### *SPECIFIC CHARACTER*

#### AND

### *SYNONYMS.*

Wings furnifhed with fmall tails. Above, brown: beneath yellowifh, with two white ftripes on the pofterior wings.

Papilio Betulæ: alis fubcaudatis fufcis fubtus luteis: pofticis ftrigis
duabus albis. *Lyn. Syft. Nat.* 2. 220,—*Fn. Sv.* 1070.
Hefperia Betulæ. *Fab. Ent. Syft.*
*Geoff. Inf.* 2. 58. 27.
*Albin. Inf. tab.* 5. *fig.* 7.
*Ernft. Pap. Europ.* 1. *tab.* 35. *fig.* 7.
*Hufnag Inf. tab* 12. *fig.* 1.
*Petiv. Gazoph. tab.* 11. *fig.* 11.

The male of this fpecies is diftinguifhed by a large fulvous mark or fpot on the anterior wings. The larva is very remarkable, being broad and flat. It is found in the months of May and June on the Alder and Sloe. Changes to Chryfalis the firft week in July. The Flies appear in Auguft.

L                    PLATE

# PLATE CCLI.

## FIG. I.

### PHALÆNA SUBERARIA.

WAVED UMBER MOTH.

LEPIDOPTERA.

*GENERIC CHARACTER.*

Antennæ taper from the bafe. Wings in general deflexed when at reft. Fly by night.

*SPECIFIC CHARACTER.*

Antennæ pectinated, yellowifh. A dark ferruginous dafh acrofs the fuperior wings, and a band of the fame on the inferior pair. The whole of the upper furface ftreaked with numerous irregular tranfverfe lines.

PHALÆNA SUBERARIA: pectinicornis lutefcens, alis fuperioribus litura, inferioribus fafcia fufco ferrugineis, omnibus ftrigofis. *Marfham MS.*

———

The Waved Umber Moth is found on the Oak in May. It is defcribed only in the MS. of T. Marfham, Efq.

L 2                    FIG.

# F I G.  II.

## PHALENA LACERTINARIA.

### *SPECIFIC CHARACTER*

#### AND

### *SYNONYMS.*

Antennæ feathered.   Wings much indentated, yellowiſh brown.
Two dark lines acroſs the anterior wings, and a light ſpot in the
middle.   Poſterior wings without any marks.

PHALÆNA LACERTINARIA, pectinicornis alis eroſis luteſcentibus:
          ſtrigis duabus punctoque medio fuſcis, poſticis imma-
          culatis.   *Lin. Syſt. Nat.* 2. 860. 204.
          *Fab. Ent. Syſt. T.* 3. *p.* 2. 135.
          *Schæf. Icon. tab.* 66. *fig.* 2. 3.
          *Degeer. Inſ.* 1. *tab.* 10. *fig.* 7. 8.
          *Reaum. Inſ.* 2. *tab.* 22. *fig.* 4—6.

Found on the Oak in May and June.

F I G.

# PLATE CCLI.

93

## F I G. III.

## PHALÆNA MACULATA.

### SPECIFIC CHARACTER.

Wings yellow ſpotted with black.

PHALÆNA MACULATA: ſeticornis alis flavis nigro maculatis. *Fab. Ent. Syſt. T.* 3. *p.* 2. 197. 244.
Phalæna Macularia. *Lynn. Syſt. Nat.* 2. 862. 213.

---

Extremely common about the hedges in the vicinity of London during the months of June and July.

J 3

PLATE

252

2

2

1

1

# PLATE CCLII.

## FIG. I. I.

### CIMEX VITTATUS.

#### GENERIC CHARACTER.

Roftrum inflected. Antennæ longer than the thorax. Wings folded crofswife. Back flat. Thorax margined. Feet formed for running.

#### SPECIFIC CHARACTER.

Black. Anterior and pofterior part of the thorax yellow: fcutellum yellow. Bafe of the antennæ, and elytra red; the latter with a longitudinal whitifh ftripe: and bent in at the apex.
CIMEX VITTATUS: thorace anterius pofteriufque fcutelloque flavis, antennarum bafi elytrifque rufis: vitta apiceque inflexo albis. *Gmel. Syft. Nat. p.* 2166. *n.* 631 ?

---

Taken on the Rofe. Fig. I. I. natural fize, and magnified.

---

## FIG. II. II.

### CIMEX POPULI.

#### SPECIFIC CHARACTER.

Oblong. Whitifh, clouded and fprinkled with brown. Antennæ fetaceous.

L 4                                                         CIMEX

CIMEX POPULI: oblongus albo fufcoque nebulofus antennis fetaceis. *Linn. Faun. Suec.* 963.

---

Very common againſt the trunks of trees, the Poplar in particular.

LINNÆAN

# LINNÆAN INDEX

## VOL VII.

### COLEOPTERA

### HEMIPTERA.

NEUROP-

3

# INDEX.

## LEPIDOPTERA.

## NEUROPTERA.

# I N D E X.

## NEUROPTERA.

---

## HYMENOPTERA.

ALPHABETICAL

# ALPHABETICAL INDEX

TO

# VOL. VII.

Flavo-

# INDEX.

8

Tecta,

# INDEX.

THE

# NATURAL HISTORY

OF

# BRITISH INSECTS;

EXPLAINING THEM

IN THEIR SEVERAL STATES,

WITH THE PERIODS OF THEIR TRANSFORMATIONS,
THEIR FOOD, OECONOMY, &c.

TOGETHER WITH THE

## HISTORY OF SUCH MINUTE INSECTS

AS REQUIRE INVESTIGATION BY THE MICROSCOPE.

THE WHOLE ILLUSTRATED BY

# COLOURED FIGURES,

DESIGNED AND EXECUTED FROM LIVING SPECIMENS.

---

By E. DONOVAN.

---

VOL. VIII.

---

LONDON:

PRINTED BY D. BYE AND H. LAW, ST. JOHN'S SQUARE, CLERKENWELL,

FOR THE AUTHOR,

And for F. and C. RIVINGTON, Nº 62, ST. PAUL'S CHURCH-YARD,

MDCCXCIX.

THE

# NATURAL HISTORY

OF

# BRITISH INSECTS.

———————

## PLATE CCLIII.

PHALÆNA PAVONIA, *minor. fem.*

EMPEROR MOTH, *female.*

The male of this fine fpecies of Phalæna has been given in the
firft Number of this Work, with a promife that the female fhould
be added in a future Plate. The larva and pupa is reprefented with
the male Infect ; but we have alfo introduced other caterpillars of
the fame fpecies with the annexed figure, to fhew the different
ftages of their growth. When young they are yellowifh : the tubercles black, with a ftripe of the fame on the fegments of the joints.
After this, the yellow bands become orange, and the tranfverfe black
ftripes appear interrupted with longitudinal bands of pale green. Some
are entirely green, except the tubercles, which are yellow, and a fmall
black fpeck on each joint ; and others are green, chequered with
black, and marked on the fide with a row of femilunar fpots. In

A 2                                                          the

the winged ſtate, we find more permanent and charaſteriſtic diſtinctions.

Linnæus, and after him Fabricius, deſcribes three varieties of
Phalæna Pavonia, α minor β media and γ major. The firſt is the
ſpecies found in this country, and in the north of Europe. The
exiſtence of the ſecond was formerly diſputed by ſome naturaliſts ;
and the laſt is ſo extremely different, at leaſt in point of magnitude,
that we may almoſt venture to remove it entirely from the two preceding.

The difference between the male and female of the common
Emperor Moth is ſtrikingly obvious ; the male is ſmaller than the
female, and the colours in general darker ; the poſterior wings alſo
are orange in the male, and not ſo in the female ; and finally, the
two ſexes may be determined by the ſtruſture of the antennæ : thoſe
of the male being nearly oval, and very deeply feathered, or peſtinated, and thoſe of the female being alſo peſtinated, but ſo ſlightly
as to appear ſetaceous. As the ſtruſture of the antennæ is an unerring criterion by which the ſexes are aſcertained, the Phalæna Pavonia
media is a phænomenon in Entomology, for both the male and female
ſo perfeſtly reſembles the *female Emperor Moth*, P. P. media, which
we have figured, that it may be miſtaken for the ſame ſpecies : the
female differs in no reſpeſt from it ; and the male only in the form
of the antennæ. We have received this remarkable ſpecies from
Italy and Germany. It is figured only by *Eſper, Phal. 3. tab. 3* ;
and is thus deſcribed by Linnæus and Fabricius, β media : " ſingu-
laris ob fœminam mari ſimillimam." The third, P. Pavonia major,
can by no means be confounded with the preceding : our ſpecimen of
it is ſix inches in breadth, and is alſo very bulky : it is found in the
*Pays de vaud.* Roeſel has given a figure of both ſexes with the larva
and pupa. The winged Inſeſt is of a dingy brown, the marks
ſomewhat ſimilar to thoſe of the common kind. The larva is large,
with the head ſmall in proportion. The whole is of a citron green,
furniſhed

# PLATE CCLIII. 5

furnished with elevated tubercles, whose summits diverge into rays like a star, and are of an azure blue colour : it is also beset with a number of long filaments of threads, each of which terminate in a little capitulum similar to the antennæ of a Butterfly.

A 3        PLATE

# PLATE CCLIV.

## FIG. I. I.

### PAPILIO PANISCUS.

*GENERIC CHARACTER.*

Antennæ clubbed at the ends. Wings in general erect when at reft. Fly by day.

*SPECIFIC CHARACTER*

AND

*SYNONYMS.*

Wings entire, divaricated ; dark brown with fulvous fpots.

HESPERIA PANISCUS: alis integerrimis divaricatis: pofticis utrinque fufcis fulvo maculatis. *Fab. Ent. Syft. T. 3. p.* 1. 328. 242.
Papilio Brontes. *Wien. Verz.* 160. 6.
Papilio Palemon. *Pall. Itin.* 1. *App. Nr.* 63.
β Papilio Silvius. *Efp. pap. tab.* 80. *fig.* 5. 6.

————————————

P. Panifcus is defcribed by Fabricius as a native of Germany, and has been lately added to the lift of Britifh Papiliones. In this country it feems a very local fpecies. It is deemed a rare Infect by Entomologifts.

A 4                           FIG.

# F I G. II.

## PAPILIO SYLVANUS.

LEPIDOPTERA.

*SPECIFIC CHARACTER*

AND

*SYNONYMS.*

Wings divaricated, brown, with fquare fpots, that appear yellow on the upper furface, whitifh beneath.

HESPERIA SYLVANUS: alis divaricatis fufcis : maculis quadratis fupra flavis fubtis albis. *Fab. Ent. Syft. T.* 3. *p.* 1. 3²6. 237.

---

Fabricius has no reference to any author for a figure of this fpecies, nor is it defcribed by Linnæus : this is the more remarkable, as the fpecies is found in great abundance in the months of May and June in this country, and is not uncommon in Sweden and Germany.

PLATE

# PLATE CCLV.

## FIG I.

### SCARABÆUS NUTANS.

#### GENERIC CHARACTER.

Antennæ terminate in a kind of club; which is longitudinally divided into lamina, two, three, or feven in number. Second joint of the anterior, or foremoft legs, armed with fpines.

#### SPECIFIC CHARACTER

AND

#### SYNONYMS.

Black; without fcutellum; anterior part of the thorax impreffed or hollow. Back of the head terminates in an erect fpine, bent at the apex.

SCARABÆUS NUTANS : exfcutellatus thorace antice impreffo, occi-
pite fpina erecta apice nutante, corpore nigro. *Fab.*
*Ent. Syft. T.* 1. *p.* 59. 194.
Scarabæus nutans.  *Oliv. Inf.* 1. 3. 145. 176. *tab.* 21. *fig.* 188.

---

A local fpecies. It is feldom found in this country. Fabricius notes it as a Saxon Infect.

FIG.

## F I G. II.

## SCARABÆUS NUCHICORNIS.

### *SPECIFIC CHARACTER*

AND

### *SYNONYMS.*

Thorax roundifh.    Back of the head armed with an erect fpine.

SCARABÆUS NUCHICORNIS: thorace rotundato, occipite fpina erecta armato. *Linn. Syft. Nat.*

SCARABÆUS NUCHICORNIS: exfcutellatus, thorace rotundato mutico, occipite fpina erecta armato, clypeo marginato. *Fab. Ent. Syft. T.* 1. 192. *p.* 58.

———————————

Not uncommon in feveral places near London.

PLATE

1

# PLATE CCLVI.

## BUPRESTIS MINUTA.

### MINUTE BUPRESTIS, or COW BURNER.

### COLEOPTERA.

### GENERIC CHARACTER.

Antennæ fetaceous, length of the thorax. Head half reɑracɑed within the thorax.

### SPECIFIC CHARACTER

### AND

### SYNONYMS.

Ovated. Wing-cafes bronged, rugged, and tranfverfely undulated with ftreaks of fine whitifh hairs.

BUPRESTIS MINUTA : elytris integris tranfverfe rugofis, thorace fubtrilobo lævi, corpore ovato nigro. *Linn. Syft. Nat.* 2. 663. 24.—*Fn. Sv.* 760.—*Fab. Ent. Syft.* 1. *p.* 2. 212. *fp.* 111.
Cucujus fufco cupreus triangularis : fafciis undulatis villofo albidis. *Geoff. Inf.* 1. 128. 6.

––––––––––––

The natural fize is reprefented at figure 1.

This pretty fpecies is found on the nut tree in May and June.

PLATE

# PLATE CCLVII.

## PHALÆNA VERBASCI.

### WATER BETONY MOTH.

### LEPIDOPTERA.

### *GENERIC CHARACTER.*

Antennæ taper from the bafe. Wings in general reflexed when at reft. Fly by night.

### *SPECIFIC CHARACTER*

### AND

### *SYNONYMS.*

Thorax crefted. Wings deflexed, margins deeply ferrated. A dark brown ftreak along the pofterior edge of the firft wings.

PHALÆNA VERBASCI: criftata alis deflexis dentato erofis: margine laterali fufco immaculato. *Linn. Syft. Nat.* 2. 850.
*Fn. Sv.* 118.
*Fab. Ent. Syft. Nat. T.* 3. *p.* 2. 120. *Sp.* 363.
*Wien. Verz.* 73. 8.
*Raj. Inf.* 168. 25.
*Geoffr. Inf.* 2. 158. 96.
*Sulz. Hift. Inf. tab.* 22. *fig.* 7.
*Reaum. Inf.* 1. *tab.* 43. *fig.* 9. 11.
*Frifch. Inf.* 6. *tab.* 9.
*Merian. Europ.* 3. *tab.* 29.

The Water Betony Moth is a very abundant fpecies; the larva feeds on the Moth Mullien, or Water Betony plant, as its name implies. It is in the larva ftate in July, becomes a pupa, and the fly is produced in April.

I

PLATE

# PLATE CCLVIII.

## PAPILIO GALATHEA.

MARBLE BUTTERFLY.

LEPIDOPTERA.

### GENERIC CHARACTER.

Antennæ clubbed at the ends. Wings erect when at reft. Fly by day.

### SPECIFIC CHARACTER

AND

### SYNONYMS.

Wings dentated, chequered with black and white, in irregular fpots and lines. Beneath, one eye-fpot on the anterior wings, and five on the pofterior ones.

PAPILIO GALATHEA: alis dentatis albo nigroque variis: fubtus anticis ocello unico, pofticis quinque. *Linn. Syft. Nat.* 2. 772. 147.
*Fab. Ent. Syft. T.* 3. *p.* 1. 239. 745.
*Schæff. Icon. tab.* 98. *fig.* 7, 8, 9.
*Roef. Inf.* 3. *tab.* 37. *fig.* 1. 2.
*Efp. pap.* 1. *tab.* 7. *fig.* 3.—*Tab.* 25. *fig.* 1.
*Ernft. Inf. Europ.* 1. *tab.* 30.
*Petiv. Muf.* 4. 3. *tab.* 1. *fig.* 1.

This

This Infect is very common in the fly state in the month of **June**. It frequents meadows, and is suppofed to feed on grafs in the larva state. The larva is very feldom met with. It is thus defcribed by fome authors: *Larva* depreffed, or flattifh, of a yellowifh colour, marked with an obfcure line down the back and fides. The *Pupa* blue, with a red tail *.

The light colour in the wings varies in different infects, fome being almoft white, and others pale yellow.

---

* *Naturf.* 14. *tab.* 2. *fig* 1.—*Fab. Ent. Syft. &c.*

PLATE

# PLATE CCLIX.

## PAPILIO SEMELE.

### BLACK-EYED MARBLED BUTTERFLY.

### LEPIDOPTERA.

### *GENERIC CHARACTER.*

Antennæ clubbed at the end. Wings erect, when at rest. Fly by day.

### *SPECIFIC CHARACTER*

### AND

### *SYNONYMS.*

Wings dentated, dark brown, with fulvous spots next to the exterior margin. Two black rings, or eyes, on the anterior, and one on the posterior wings.

PAPILIO SEMELE, alis dentatis: fascia maculari fulva ocellisque duobus: anticis subtus disco baseos. *Linn. Syst. Nat.* 2. 772. 148.—*Fn. Sv.* 1051.
*Fab. Ent. Syst. T.* 3. *p.* 1. 232. 725.
*Esp. pap.* 1. *tab.* 8. *fig.* 1.
*Schæff. Icon. tab.* 207. *fig.* 3, 4.
*Sulz. Hist. Inf. tab.* 17. *fig.* 5, 6.

This species lives chiefly in the woods. It appears in the winged state in July. The larva is not clearly ascertained by any author.

B                    PLATE

# PLATE CCLX.

## PAPILIO ATALANTA.

### RED ADMIRABLE BUTTERFLY.

### LEPIDOPTERA.

### GENERIC CHARACTER.

Antennæ clubbed at the ends. Wings erect when at reft. Fly by day.

### SPECIFIC CHARACTER

#### AND

### SYNONYMS.

Wings dentated, black with white fpots. A red ftripe acrofs the anterior wings, and another along the pofterior margin of the lower wings.

PAPILIO ATALANTA: alis dentatis nigris albo maculatis: fafcia communi purpurea anticarum utrinque pofticarum marginali. *Linn. Syft. Nat.* 2. 779. 175.—*Fn. Sv.* 1060.—*Fab. Ent. Syft. T.* 3. *p.* 1. 118. 362.
*Albin. Inf.* 3. *fig.* 4.
*Degeer Inf.* 1. *tab.* 22. *fig.* 5.
*Roef. Inf.* 1. *pap.* 1. *tab.* 6.
*Sepp. Inf.* 1. *tab.* 1.
*Schæff. Icon. tab.* 148. *fig.* 1, 2.
*Ernft. Inf. Europ.* 1. *tab.* 6.
*Geoff. Inf.* 2. 40. 6.
*Ammer Inf. tab.* 24.

The

The red admirable Butterfly is certainly a very common fpecies, but as one of the moft beautiful this country can boaft of, is entitled to our particular confideration.

The Caterpillars are of feveral kinds, according to the different ftages of growth. In the laft fkin they are green, with a yellow ftripe on each fide of the belly, and befet with curioufly ramified, or branching fpines: fometimes they are black, with a yellow belly, or black, variegated with red, brown, and yellow, The Chryfalis is of a dark colour, ornamented with feveral fpots of fhining gold. The Caterpillars are found on the nettle in June and July: it remains in Chryfalis twenty-one days, and the Butterfly appears in Auguft. There are two broods of this fpecies in general every feafon.

PLATE

# 261

# PLATE CCLXI.

### F I G. I. I.

## HIPPOBOSCA EQUINA.

#### HORSE, OR SPIDER FLY.

#### DIPTERA.

### GENERIC CHARACTER.

Roftrum bivalve, cylindrical, obtufe and wavering. Antennæ fhort, fetaceous. Without ftemmata. Feet armed with many nails or crotchets.

### SPECIFIC CHARACTER

#### AND

### SYNONYMS.

Wings obtufe. Thorax variegated. Feet armed with three crotchets.

HIPPOBOSCA EQUINA, alis obtufis, thorace albo variegato, pedibus tetradactylis. *Linn. Syft. Nat.* **2.** 1010. **I.** *Fn. Sv.* 1921.—*Fab. Ent. Syft. Vol. 4. p.* 415. **I.** *Degeer. Inf.* 6. 257. **I.** *tab.* 16. *fig.* **I.** *Reaum. Inf.* 2. *tab.* 179. *fig.* 8, 9.

Few fpecies of Hippobofcæ have hitherto been difcovered. Four kinds were known to Linnæus, and Fabricius has not encreafed

C

that

that number in either of his Entomological Syftems. A fifth fpecies is defcribed by Gmelin in the laft edition of the Syftema Naturæ, under the name of Uralenfis *. The fpecies in the Entomologia Syftematica of Fabricius, are *Equina, Avicularia, Hirundinis,* and *Lovina.* The firft is rather larger than the others, and is well known by the common name of Horfe Fly, becaufe it frequently molefts thofe animals, and attaching itfelf to their bodies, penetrates the fkin, and fucks their blood. The three other fpecies are alfo natives of this country, and like the H. equina, fubfifts on the blood of certain animals. The H. avicularia is found chiefly on the bodies of birds, and H. hirudinis more efpecially on fwallows. Thefe creatures are all of a difgufting form, flat, and hard : they adhere very tenacioufly by means of the nails or crotchets of their talons, which in this genus are numerous ; and are not eafily killed by preffure. The H. equina has three fharp incurvated nails to each foot, the H. hirundinis is furnifhed with twice that number on each.

F I G.  II.

HIPPOBOSCA  AVICULARIA.

*SPECIFIC  CHARACTER*

AND

*SYNONYMS.*

Wings obtufe. Thorax without fpots, and of one colour.

* Hippobofca uralenfis : atra hirfuta dorfo ordinibus tribus veficularum albarum nitentium. Lepechin it. 1. t. 19. f. 9. *habitat in deferto Uralenfi.*

HIPPOBOSCA

# PLATE CCLXI. 23

HIPPOBOSCA AVICULARIA, alis obtufis, thorace unicolore. *Linn. Syft. Nat.* 2. 1010. 2.—*Fn. Sv.* 1922.—*Fab. Ent. Syft.* 4. *p.* 415. 2.

---

The figure of Hippobofca avicularia in the works of Sulzer nearly agrees with our infect, except in colour; it is of a dull brown, with lefs of the green caft.

C 2

PLATE

# 262

# PLATE CCLXII.

## FIG. I. I.

## PHALÆNA ATRIPLICIS.

### WILD ARRACH MOTH.

### LEPIDOPTERA.

### *GENERIC CHARACTER.*

Antennæ taper from the bafe, wings in general deflexed when at reft.   Fly by night.

### *SPECIFIC CHARACTER*

### AND

### *SYNONYMS.*

Thorax crefted, anterior wings, brown, clouded and marked with undulated ftreaks, and a two-cleft or forked yellow mark in the middle.

PHALÆNA ATRIPLICIS, criftata alis deflexis: anticis fufco nebulofis; litura media flava bifida.   *Lin. Syft. Nat.* 2.
854. 173. *Fab. Ent. Syft. T.* 3. *p.* 95. *fp.* 282.
*Roef. Inf.* 1. *phal.* 2. *tab.* 31.

———————

This fpecies is noticed by Harris and Berkenhout, as a native of Great Britain, under the trivial name of *Wild arrach Moth*; we have ever confidered it a local fpecies.   It is found in Cambridgefhire,

C 3

the

the Caterpillar in Auguſt, the Fly in September. The wild orach and common dock are its favourite food.

---

## FIG. II. II.

## PHALÆNA UMBRATICA.

### Shark Moth.

### *SPECIFIC CHARACTER*

#### AND

#### *SYNONYMS.*

Thorax creſted. Wings deflexed, lanceolated, greyiſh, ſtriated longitudinally with pale black. A faint reddiſh daſh in the middle, marked with two black ſpots.

Phalæna Umbratica: criſtata alis deflexis ſtriatis lanceolatis canis: macula centrali ferruginea ; punctis duobus nigris. *Linn. Syſt. Nat.* 849. 150. *Fn. Sv.* 1184.—*Fab. Ent. Syſt. T. 3. p.* 122. 368.
Noctua lucifuga. *Wien. Verz.* 312. 11 ?
*Roeſ. Inſ.* 1. *phal.* 2. *tab.* 25.

---

The larva of Phalæna Umbratica is ſometimes of a dirty brown, with ſpots of clay colour ; it feeds on the ſow thiſtle ; changes into the Chryſalis ſtate in May, and the Flies appear in June.

PLATE

# 263

1

2

# PLATE CCLXIII.

## FIG. I.

## PHALÆNA PINETELLA.

### Pearl Veneer Moth.

### *GENERIC CHARACTER.*

Antennæ, taper from the bafe. Wings in general deflexed, when at reft. Fly by night.

### *SPECIFIC CHARACTER*

#### AND

### *SYNONYMS.*

Firft wings yellowifh brown, with two large whitifh, or pearl-like fpots on each.

Tinea Pinetella : alis anticis flavis : maculis duabus albiffimis, anteriore oblonga, pofteriore ovata. *Linn. Syft. Nat.* 2. 886. 356.—*Fn. Sv.* 1368. *Fab. Ent. Syft. T.* 3. *p.* 2. 294. *Clerk. phal. tab.* 4. *fig.* 7. *Panz. Faun. Germ.* 6. *tab.* 22. *Wien Verz.* 134. 7.

———————

We met with a fpecimen of this beautiful little Moth, in June 1798, in Norwood. It is certainly a rare and interefting Britifh

C 4                                              fpecies;

ſpecies; though long ſince known to collectors of Inſects by the trivial Engliſh name above adopted. It may not be improper to add, that this ſpecimen was found in a willow tree, as thoſe continental Authors who have noticed it, ſay it inhabits pine trees.

---

## FIG. II.

## PHALÆNA COLONELLA.

### SPECIFIC CHARACTER

#### AND

### SYNONYMS.

Firſt wings oblong, greyiſh, with two black ſpots in the centre. A faint undulated bar acroſs the interior, and another near the exterior part of each wing.

TINEA COLONELLA: alis oblongis cinereis: punctis duabus atris ante ſtrigam curvam undulatam obſoletam. *Linn. Syſt. Nat.* 2. 883. 346. *Fn. Sv.* 1358.—*Fab. Ent. Syſt. T.* 3. *p.* 2. *p.* 288. 5.

---

Found on the alder in July.—*Combe wood, Surrey.*

PLATE

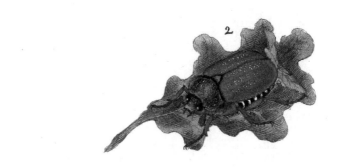

# 264

# PLATE CCLXIV.

## FIG I.

## LUCANUS PARALELEPIPEDUS.

### SMALL BLACK STAG BEETLE.

### *GENERIC CHARACTER.*

Antennæ terminate in a club or knob compreſſed on one ſide, and divided into laminæ or teeth. Maxillæ or jaws, ſtrong, porreĉted before the head, and armed with teeth.

### *SPECIFIC CHARACTER*

### AND

### *SYNONYMS.*

Black : body depreſſed. Maxillæ furniſhed with a large elevated tooth on the interior ſide.

LUCANUS PARALELEPIPEDUS: mandibulis dente laterali elevato, corpore depreſſo. *Linn. Syſt. Nat.* 2. 561. 6. *Fab. Ent. Syſt. T.* 1. *p.* 2. 239. 11.

*Platycerus* niger, elytris lævibus, capitis punĉto duplici prominente. *Geoff. Inſ.* 1. 62. 3.
*Degeer. Inſ.* 4. *tab.* 12. *fig.* 9.
*Schæff. Icon. tab.* 63. *fig.* 7.|
—— *Elem. tab.* 101. *fig.* 1.

Three

Three fpecies of Lucani are found in this country. The largeft is Lucanus Cervus, of which a figure has been given in the early part of this Work. The prefent fpecies and L. caraboides complete the lift of Britifh Lucani.

The firft kind is known to live in the larva ftate in decayed wood, and it is fuppofed the fame habits of life belong to others of this tribe. Lucanus Paralelepipedus ranks amongft the rareft Coleopterous infects of this country; in Denmark and Germany it is not uncommon.

---

## F I G. II.

### SCARABÆUS MELOLONTHA.

#### TREE BEETLE, or COCKCHAFER.

*GENERIC CHARACTER.*

Antennæ terminate in a kind of club, which is longitudinally divided into laminæ. Second joint of the anterior legs, armed with fpines.

*SPECIFIC CHARACTER*

AND

*SYNONYMS.*

Furnifhed with fcutellum. Brown. Thorax hairy. Tail bent or inflected, a triangular white fpot at each incifure of the abdomen.

SCARABÆUS MELOLONTHA: fcutellatus muticus teftaceus, thorace villofa, cauda inflexa, incifuris abdominis albis. *Linn. Syft. Nat.* 2. 554. 60.—*Fn. Sv.* 392.

Melolontha

# PLATE CCLXIV. 31

Melolontha vulgaris. *Roef. Inf.* 2. *Scarab.* 1. *tab.* 1.
Melolontha vulgaris. *Fab. Spec. Inf.* 1. *p.* 35. 3.—*Syft. Ent. T.* 1.
     *p.* 2. *p.* 155. 3.
     *Degeer. Inf.* 4. *tab.* 10. *fig.* 14.
     *Voet. Scarab. tab.* 6. *fig.* 45. 46.

---

The Cockchafer is found in the greateft abundance throughout Europe, and in fome countries are infinitely more detrimental to agriculture than in England. In the larva ftate, it lies buried beneath the furface of the earth, where it does incredible mifchief by devouring the roots of plants. After remaining three years in the larva ftate it becomes a winged Infect, and leaving its fubterraneous habitation, is not lefs detrimental to the corn and other kinds of grain, and alfo to trees; particularly the Oak. In fome feafons, when they are numerous, we find the Oak ftripped of its leaves, and otherwife much injured by thefe voracious animals. About noon the Cock-chafers collect in confiderable multitudes among the leaves of the lower boughs of the Oak, when they may be eafily taken, by fhaking or beating them into bafkets, &c.

Several varieties of this fpecies are mentioned by authors; parti-cularly one with a red thorax, by Roefel.

---

# F I G. III.

## SCARABÆUS STERCORARIUS.

### COMMON DOR, or CLOCK BEETLE.

*Both thorax and head without fpines.

*SPECIFIC*

## PLATE CCLXIV.

### *SPECIFIC CHARACTER.*

Furniſhed with ſcutellum. Black: clypeus rhombic with a ſmall elevation or protuberance in the middle. Wing caſes furrowed.

SCARABÆUS STERCORARIUS: ſcutellatus muticus ater, clypeo rhombeo, vertice prominulo, elytris ſulcatis. *Linn. Syſt. Nat.* 2. 551. 42.
*Fn. Sv.* 388.
*Fab. Syſt. Ent.* 17. 60.
*Spec. Inf.* 1. *p.* 18. 74.
*Schæff. Icon. tab.* 23. *fig.* 9.

———————————

Like the preceding ſpecies, this Beetle lives in the larva ſtate in the earth. The Winged Infeⅽt is found in the dung of animals, carrion, &c. Few infeⅽts are found more numerous or general throughout Europe than this. The colour is black, with very vivid gloſſes of ſhining blue, green, and purple, in different ſpecimens; and we have a variety of it from the ſouth of Europe that is entirely of a dull reddiſh colour.

PLATE

# PLATE CCLXV.

## PHALÆNA JOTA.

GOLDEN *i* MOTH.

LEPIDOPTERA.

### GENERIC CHARACTER.

Antennæ taper from the bafe. Wings in general deflexed when at reft. Fly by night.

### SPECIFIC CHARACTER

AND

### SYNONYMS.

Thorax crefted. Anterior wings ferruginous red, varioufly fhaded, and marked in the centre with an inverted golden *i*.

PHALÆNA IOTA: criftata alis deflexis, anticis ferrugineo grifeis I
    refupinato aureo infcriptis. *Fab. Syft. Ent.* 608. 76.
    —*Sp. Inf. T.* 2. *p.* 228. 98.—*Ent. Syft. T.* 3. *p.* 81.
    *Sp.* 237.
PHALÆNA IOTA: alis primoribus ferrugineo-grifeis, I refupinato
    aureo infcriptis. *Gmel. Syft. Nat.*—*Ent. p.* 2557. 130.
    *Roef. Inf.* 1. *phal.* 3. *tab.* 5.

———————

Linnæus very accurately defcribed this Infect under the fpecific name *Iota* in the early edition of the Syftema Naturæ. It is given as a native of Germany, and without a reference to any figure. Fabricius has followed the fame defcription throughout all his works,

without

without noticing any figure of it alfo, until the Entomologia Syftematica appeared, when a reference is given to *tab*. 5. *Phal*. 3. *Vol*. I. of
Roefel's Infects.   This is however erroneous, for Linnæus includes
that reference among the Synonyms of his *Phalæna Gamma,* and
*Gmelin*, his lateft editor, continues the fame amongft the Synonyms of
Gamma likewife.   Kleman, who revifed the laft edition of Roefel's
work, is decidedly of opinion, that the figure is that of Gamma,
and affigns the Linnæan fpecific name to it accordingly.   Whatever
reafon influenced the opinion of Fabricius, that the figure in queftion
was the true Phalæna Iota of Linnæus, when that author himfelf
fays otherwife, is not mentioned by Fabricius, and is by no means
clear to us; but one circumftance cannot efcape obfervation: Fabricius has in no inftance referred to the figure in Roefel's plate before the publication of his laft work, though all the other Synonyms
of *Phalæna Gamma* are the fame in every edition of the Fabrician
fyftems as in thofe of his predeceffor, Linnæus.

This Infect is very beautiful, and much rarer than the following
fpecies.   It is found on the common, and white dead nettles, or
archangel.   The larva has twelve feet; is without hairs, green;
and fpotted with white.   The Moth appears in June.

---

## F I G. II.

### PHALÆNA GAMMA.

COMMON γ MOTH.

*SPECIFIC CHARACTER.*

AND

*SYNONYMS.*

Thorax crefted.   Anterior wings brown, with a golden γ infcribed
in the centre.

PHALÆNA

# PLATE CCLXV.     35

PHALÆNA GAMMA: criſtata alis deflexis dentatis: anticis fuſcis γ
    aureo inſcriptis.    *Linn. Syſt. Nat.* 2. 843. *p.* 127.
      *Fab. Ent. Syſt. T.* 3. *p.* 79. *ſp.* 228.

PHALÆNA GAMMA: alis primoribus fuſcis γ aureo inſcripts
      *Gmel. Linn. Syſt. Nat.—Ent.* 2555. *ſp.* 127.
      *Geoff. Inſ.* 2. 156. 92.
      *Goed. Inſ.* 2. *t.* 21.
      *Rag. Inſ. p.* 163. *n.* 16.
      *Petiv. Gazoph.* 4. 6.
      *Schæff. Icon. tab.* 84. *fig.* 5.
      *Friſch. Inſ.* 5. 15.
      *Reaum. Inſ.* 2. *tab.* 26. *fig.* 5.
      *Albin. Inſ. tab.* 79. *fig.* G. H.
      *Sepp. Inſ.* 5. *tab.* 1. 61.

---

    Phalæna Gamma is one of thoſe ſpecies which feed indifferently
on many kinds of plant. It is often found amongſt nettles and
other low herbage, or in gardens amongſt cabbages, &c. It is of
a green colour, with pale or whitiſh longitudinal ſtripes on the back,
and the ſides yellow. The Moth is found in Auguſt and September.

PLATE

[ 37 ]

# PLATE CCLXVI.

## FIG. I, II.

### PHALÆNA LEMNATA

SMALL WHITE CHINA MARK MOTH.

LEPIDOPTERA.

### GENERIC CHARACTER.

Antennæ taper from the bafe. Wings in general deflexed when at reft. Fly by night.

### SPECIFIC CHARACTER

AND

### SYNONYMS.

Antennæ fetaceous. Wings fnowy white. A black ftreak next the pofterior margin of the lower pair, marked with four white fpots.

PHALÆNA LEMNATA : feticornis alis niveis: pofticis fafcia terminali nigra ; punctis quatuor albis. *Lynn. Syft. Nat.* 2.
874. 278.—*Fn. Sv.* 1301.—*Fab. Ent. Syft. T.* 3. *p.* 2.
*p.* 215. *fp.* 319.
*Raj. Inf.* 205. 102.
*Reaum. Inf.* 2. *tab.* 12. *fig.* 14, 15.

---

Linnæus and Fabricius defcribe only the male of this fpecies; the female is rather larger, the colour pale brown, with markings

D

fimilar

fimilar to thofe of the other fex, but of a reddifh colour. The characteriftic black line, with white fpots on the pofterior wing, is the fame as in the male Infect.

This Moth derives its fpecific name, Lemnata, from the food of its Caterpillar, or larva, which is ufually fome fpecies of *lemna*, (duckweed) or other aquatic plant. In the winged ftate it is a very common Infect, particularly in marfhy ground, and the fides of ponds and ditches. In general we find two broods of them in the fummer; the firft appears in May, the fecond in July or Auguft.

---

## F I G. III.

### PHALÆNA DECUSSATA.

#### PRETTY WIDOW MOTH.

#### GEOMETRA.

### *SPECIFIC CHARACTER.*

Antennæ fetaceous. Wings cinereous, with four black ftreaks; the two middle ones of a ferpentine form, and croffing each other. Several undulated dotted black lines.

PHALÆNA DECUSSATA: felicornis, alis cinereis; ftrigis quatuor, mediis decuffatim flexuofis, lineolis undulatis atomif-que nigris.

---

A rare and new fpecies. It has been taken at Faverfham by Mr. Crewe, and trivially named the Pretty Widow Moth.

FIG.

PLATE CCLXVI.

39

# FIG. IV.

## PHALÆNA ATRALIS.

### SPECIFIC CHARACTER

AND

### SYNONYMS.

Wings black, with two white fpots on each.

PHALÆNA ATRALIS: alis atris: maculis duabus niveis.   *Linn.*
        *Mant.* 540.—*Fab. Syft. Ent. T.* 3. *p.* 2. 241. *fp.* 422.
Phalæna funera.   *Myll. Zool. Dan.* 132. 1524.
Phalæna guttalis.   *Wien. Verz.* 124. 45.
Phalæna funebris.   *Act. Nidrof.* 4. *tab.* 16. *fig.* 17.

———————————

Taken in June, on Epping Foreft.

PLATE

# PLATE CCLXVII.

## FIG. I. II.

### PHALÆNA DEGEERELLA.

LONG HORN JAPANNED MOTH.

LEPIDOPTERA.

*GENERIC CHARACTER.*

Antennæ, taper from the bafe. Wings in general deflexed, when at reft. Fly by night.

TINEA.

*SPECIFIC CHARACTER*

AND

*SYNONYMS.*

Black, bronzed, or changeable to gold. A yellow indented band acrofs the anterior wings. Antennæ very long.

PHALÆNA DEGEERELLA: alis atro aureis: fafcia flava, antennis
       longis. *Linn. Syft. Nat.* 2. 895. 426.—*Fn. Sv.* 1393.
ALUCITA DEGEERELLA. *Fab. Ent. Syft. T.* 3. *p.* 2. *p.* 341. 40.

———————

The two fexes of this beautiful and extraordinary little Infect has been defcribed as two diftinct fpecies. They differ principally in the form and fize of the antennæ ; in the female thefe are entirely fetaceous, or like a fine hair, three times the length of the whole body: the antennæ in the male are fhorter, and are thick in the middle, not very unlike thofe of fome kinds of Sphinges. It is not uncommon in hedges in May and June.

E

FIG.

# PLATE CCLXVII.

## FIG. III. III.

## PHALÆNA PODAELLA.

SCARCE JAPANNED MOTH.

LEPIDOPTERA.

TINEA.

*SPECIFIC CHARACTER*

AND

*SYNONYMS.*

Antennæ very long. Wings golden black. A narrow gold or yellow line of equal breadth acrofs the anterior wings.

PHALÆNA PODELLA : antennis longiffimis, alis nigro-æneis, anticis fafciâ equali angufta aurea.

PHALÆNA PODELLA : antennis mediocribus alis nigris fafcia albida. *Linn. Syft. Nat.* 896. 428.

Alucita Podaella. *Fab. Ent. Syft.* 3. *p.* 2. 341. *n.* 42.

---

Like the preceding fpecies, the two fexes of PHALÆNA *Tinea* PODAELLA are diftinguifhed by the different form of the antennæ ; and Linnæus has evidently defcribed only the male, or he would not have defined it—" *Antennis mediocribus,*" for the antennæ of the female are remarkably long.

The

PLATE CCLXVII. 43

The reference made by Fabricius to *Geoff. Inf.* 2. 194. 32. for this Infect, is certainly erroneous; for the Infect therein defcribed is ftated to have a white band acrofs the four wings, a white point near the exterior margin of the upper wings, and a fmall tranfverfe white mark towards the bafe. Linnæus does not quote *Geoffroy* for this Infect; and it is evident, Fabricius has confined his reading to the Latin fpecific defcription, which certainly agrees with the Linnæan defcription of Podaella in the *Syft. Nat.*

The natural fize and magnified appearance is fhewn at Fig. III. III.

E 2

PLATE

1

2

1

※
2

# PLATE CCLXVIII.

## FIG I.

## HIPPOBOSCA HIRUNDINIS.

### *GENERIC CHARACTER.*

Roſtrum bivalve, wavering. Feet armed, with many nails.

### *SPECIFIC CHARACTER*

AND

### *S Y N O N Y M S.*

Wings taper to a point from the baſe. Six claws on each foot.

HIPPOBOSCA HIRUNDINIS: alis ſubulatis, pedibus hexadactylis.
*Linn. Syſt. Nat.* 2. 1010. 3. *Fn. Sv.* 1923.
*Geoff. Inſ.* 2. 547. 2.—*Panz-Faun. Inſ. Germ.*

———————

Found on the bodies and neſts of ſwallows.

The ſmalleſt Figure denotes the natural ſize.

E 3           FIG.

## FIG. II.

### HIPPOBOSCA OVINA.

#### APTEROUS.

### *SPECIFIC CHARACTER.*

Without wings.

HIPPOBOSCA OVINA: alis nullis. *Linn. Syft. Nat.* 2. 1011. 4.
*Fn. Sv.* 1924.
*Frifch. Inf.* 5. *tab.* 18.

---

  This is the fourth and laſt ſpecies of the Hippoboſcæ found in Great Britain.

PLATE

# 269

# PLATE CCLXIX.

## SPHINX OCELLATA.

### EYED HAWK MOTH.

### *GENERIC CHARACTER.*

Antennæ thickeſt in the middle. Wings deflexed when at reſt. Fly by night.

### *SPECIFIC CHARACTER*

### AND

### *SYNONYMS.*

Wings angulated. Poſterior pair red, with a large blue eye in the middle of each.

SPHINX OCELLATA: alis angulatis, poſticis rufis ocello cœruleo. *Fab. Syſt. Ent.* 536. 1.

SPHINX OCELLATA, alis repandis, poſticis ocellatis. *Linn. Syſt. Nat.* 2. 796. 1.—*Fn. Sv.* 1083.

*Phalæna* alis inferioribus macula ophthalmoide inſignibus. *Alb. Inſ. tab.* 8. *fig.* 2.

*Drury Inſ.* 2. *tab.* 25. *fig.* 2. 3.

*Roeſ. Inſ. phal.* 1. *tab.* 1.

*Schæff. Icon. tab.* 99. *fig.* 5. 6.

*Merian Europ.* 2. *tab.* 87.

---

The Sphinges are only, in a few inſtances, remarkable for that gaiety and ſplendour of colours, which render the Butterfly tribe ſo

E 4                                        pleaſing

pleafing and interefting to general obfervers. There is, however, a peculiar grace and elegance of form throughout the Sphinges, which immediately diftinguifh them from the Phalænæ, or third tribe of Lepidopterous Infects ; and their colours, though chafte or obfcure, are, for the moft part, very prettily diverfified. The Sphinges of Great Britain are not numerous ; and, in general, the more beautiful kinds are rare. The Sphinx Ocellata is certainly an exception to fuch remark, for we have not a finer or more abundant fpecies of the tribe in this country. It is alfo no lefs common in other parts of Europe ; and few authors, who have treated on the Entomology of any country in Europe, have neglected to give it a place in their works.

The larva of the Sphinx Ocellata is found on the willow, in May, June, and July, when it becomes a pupa, and remains in the earth till June following ; and then comes forth in the perfect ftate.

P L A T E

# PLATE CCLXX.

## GRYLLUS MIGRATORIUS.

### MIGRATORY LOCUST.

*GENERIC CHARACTER.*

Head infleted, armed with jaws, and furnifhed with palpi. Wings wrapped round the fides of the body, and concealed under the elytra. Feet armed, with two nails. Pofterior legs formed for leaping.

*SPECIFIC CHARACTER*

AND

*SYNONYMS.*

Thorax of one fegment, and fomewhat keeled, or rifing in a lon-gitudinal line, in the middle ; mandibules blue.

GRYLLUS MIGRATORIUS : thorace fubcarinato : fegmento unico, mandibulis cœruleis. *Linn. Syft. Nat.* 2. 700. 41.—
*Fn. Sv.* 871.
*Roef. Inf.* 2. *Gryll. tab.* 24.
*Edw. birds.* 208. *tab.* 208.
*Degeer. Inf.* 3. 446. 1. *tab.* 23. *fig.* 1.
*Seb. Muf.* 4. *tab.* 65. *fig.* 21.

---

The Gryllus Migratorius, or Migratory Locuft, has always been claffed amongft the Infects of this country on local authority. It certainly vifited Great Britain in the year 1748, and feveral fmall flocks of them were feen in the environs of London, where they

caufed

caufed much confternation, according to the authors of that time: M. Edwards in particular, gave a figure, and a fhort account of it in his Hiftory of Birds, [Plate 208] under the name of Great Brown Locuft. It is not certain that they have appeared, at leaft in confiderable numbers in this country, fince that period; and it is probable, if any fpecimens of them were then collected, they are now loft; for we have not hitherto been able to afcertain, pre- cifely, an Englifh fpecimen of it in any cabinet, and this confider- ation alone, has induced us to defer adding this remarkable creature to our illuftration of Britifh Infects.

In the month of September, 1799, LADY AYLESFORD moft obligingly communicated a living fpecimen of it to us, and we deem ourfelves particularly fortunate in the opportunity it affords us to afcertain the difference, however inconfiderable, between the variety found in England and thofe of warmer countries. It is reprefented in the annexed plate, both in a refting pofition and with the wings expanded. This fpecimen was found in a barley field near Pack- ington in Warwickfhire. It lived feveral days after its arrival in London, and would probably have furvived much longer had it not been injured in the journey, and weakened by long confinement. We remarked, that in feeding, it cuts the ftalk afunder in the middle, or near the root, and tearing off the leaves, eats only the pith; this may, in fome meafure, account for the great mifchief and depredation thefe creatures commit, when they fettle in vaft numbers on any tract of cultivated land.

Though we regard the Gryllus Migratorius as an object of cu- riofity in this country, in many others they are the terror of the inhabi- tants. We are not, perhaps, to admit implicitly the relations of all authors, but in thofe of credibility and information, we find abundant reafon to hope it may ever remain, as at prefent, a rare Britifh fpecies. It is faid to be very numerous in Tartary, from whence at certain periods it migrates weftward, and vifits the fouth of Europe in incredible quantities. Of all the authors who have treated on the hiftory of this Infect, none are more fatisfactory than Roefel;

his

I

# PLATE CCLXX. 51

his obfervations are given in detail, but they are the refult of ufeful information, and may be deemed a proper fequel to our concife account of it; when we confider that his obfervations relate to the fame event and time, the legions of Locufts that appeared in this country, and throughout Europe in 1748.

\* " I have already," fays Roefel, " given particulars of thofe dangerous guefts in the beginning of the year 1749, and at the fame time added figures of them †. I fhall now only obferve, that in the courfe of the fame year, they not only appeared in Poland, Hungary, Auftria, Bohemia, Silefia, Bavaria, but alfo in Franconia, and confequently in the environs of our city. Thanks be to God, they have not been numerous about our city, but from Windfheim we have intelligence of the third of September, that they had frightened the inhabitants with their legions, but they departed again on the fourth; fince then, a much refpected patron has kindly fent me, the IMPERIAL AND ROYAL HUNGARIAN EDICT of the prefent year 1749, together with a printed defcription of the Infect, and proper inftructions to the imperial fubjects, how to extirpate them, and I thought it neceffary, as they contain much good and ufeful matter, to infert them." Then he proceeds with *Befchreibung Deren Anno* 1747, &c. &c. or a defcription of the Locuft, as given in the Imperial Edict, &c. In the courfe of which, and the defcription in page 145, we are informed, that this Locuft lives three years. The female depofits her eggs in a kind of bag; the eggs are about a quarter of an inch in length: of a flender oblong form, and placed within the bag lengthwife, fo as to form four or five tiers. It is moft advifed to dig for them in this ftate, and burn them; or if they fhould alight in the winged ftate, as it is known that the noife of bells, &c. will allure them to any particular fpot, it is recommended to form deep ditches, and decoy, drive, or beat them into the water in multitudes.

---

\* Under the fection *Der Heufchrecten und Grillenfammlung*, &c. &c. Vol. II. p. 103.

† Referring to his plate 24, of LOCUSTA GERMANICA and defcription of the Infects annexed.

PLATE

# PLATE CCLXXI.

## PAPILIO HYPERANTHUS,

### RINGLET BUTTERFLY.

### GENERIC CHARACTER.

Antennæ clubbed at the end. Wings erect when at reft. Fly by day.

### SPECIFIC CHARACTER

#### AND

### SYNONYMS.

Wings entire, brown. Beneath, three eyes or rings on the anterior, and five on the pofterior wings.

PAPILIO HYPERANTHUS: alis integerrimis fufcis fubtus anticis
        ocellis tribus, pofticis duobus tribufque. *Linn. Syft.
        Nat.* 2. 768. 127.—*Fn. Sv.* 1043.—*Fab. Ent. Syft.*
        T. 3. p. 1. p. 216. fp. 677.
        *Degeer. Inf.* 2. tab. 2. fig. 9. 10.
        *Schæff. Icon.* tab. 127. fig. 1, 2.
        *Efp. Pap.* 1. tab. 5. fig. 1.
        *Naturf.* 8. tab. 3. fig. D.

The larva of this Butterfly is very rarely met with; it feeds on the roots of grafs; is hairy, of an afh colour, with a black line down the fide, and the tail forked. The pupa is gibbous, brown, and fpotted with yellow.

Papilio

Papilio Hyperanthus is found in the winged ſtate in June and July. It frequents lanes, and hedges on dry and elevated banks, ſuch as are common in the ſandy and chalky ſoils of Kent.

P L A T E

272

# PLATE CCLXXII.

## PHALÆNA FURCULA.

### KITTEN MOTH.

#### GENERIC CHARACTER.

Antennæ taper from the bafe. Wings in general deflexed when at reft. Fly by night.

##### BOMBYX.

#### SPECIFIC CHARACTER

AND

#### SYNONYMS.

Thorax variegated. Anterior wings grey, fprinkled with black : bafe and apex white, with black fpots. Pofterior wings white, with a marginal row of black points.

PHALÆNA FURCULA : thorace variegato, alis grifeis bafi apiceque albis nigro punctatis. *Linn. Syft. Nat.* 2. 823. 51.— *Fn. Sv.* 1122.—*Fab. Ent. Syft. T.* 3. *p.* 1. 475. *fp.* 213. *Panz. Faun. Inf. Germ.* 4. *tab.* 20. *Wilks. pap.* 13. *tab.* 1. *fig.* 1. *Sepp. Inf.* 4. 29. *tab.* 6.

Phalæna Furcula is a neat and interefting Britifh Infect. Except in fize it is very fimilar to Phalæna Vinula; and the Aurelians, from this fimilarity, and a fanciful reference to the brindled appearance

of

of the anterior wings, have whimfically, but not inaptly, given their Englifh trivial names: Phalæna Vinula, being the largeft, is called the Pufs Moth; and Phalæna Furcula, the Kitten. The laft is rare: Phalæna Vinula very common.

The larva is found on the willow in July. It remains in the pupa ftate the whole winter : the Moth comes forth in May.

PLATE

273

1

# PLATE CCLXXIII.

## CHRYSOMELA NITIDULA.

### GENERIC CHARACTER.

Antennæ compofed of globular articulations, which encreafe in bulk towards the end. Thorax and elytro without margin.

### SPECIFIC CHARACTER

### AND

### SYNONYMS.

Feet formed for leaping. Wing's cafes fhining green. Head and thorax, crimfon, refplendent with gold. Legs ferruginous.

CHRYSOMELA NITIDULA: faltatoria, elytris cœruleis, capite thoraceque aureo. *Faun. Suec.* 542.

GALLERUCA NITIDULA, faltatoria viridis nitens, capite thoraceque aureis, pedibus ferrugineis. *Fab. Ent. Syft.* 1. 30. *fp.* 81.

——————————

This beautiful little fpecies is fometimes found on the willow and alder. The fmalleft figure denotes the natural fize.

F                    PLATE

# PLATE CCLXXIV.

## FIG. I. II.

### PHALÆNA HUMULI.

#### GHOST MOTH.

*GENERIC CHARACTER.*

Antennæ taper from the bafe. Wings in general deflexed when at reft. Fly by night.

*SPECIFIC CHARACTER*

AND

*SYNONYMS.*

Female yellow, with fulvous marks. Male fnowy white.

PHALÆNA HUMULI : alis flavis fulvo ftriatis maris niveis. *Linn. Syft. Nat.* 2. 833. 84.—*Fn. Sv.* 1147.
HEPIALUS HUMULI : *Fab. Ent. Syft. T.* 3. *p.* 2. 5. *fp.* 1.
*Degeer. Inf.* 1. *tab.* 7. *fig.* 5, 6.
*Sulz. Hift. Inf. tab.* 22. *fig.* 1.

---

The male and female of *Phalæna Humuli* are very diffimilar, and may eafily be miftaken for diftinct fpecies. The male is perfectly white, with a glofs like fatin, the abdomen, antennæ, and margin of the wings excepted, for thefe are reddifh brown. The female is of a fine yellow colour, with feveral fulvous or orange marks; and is fomewhat larger than the other fex.

The larva lives in the earth, at the roots of the Burdock and hop. It is of a very pale or whitifh colour, with a brown head, and fixteen feet.

FIG.

## F I G. III.

### PHALÆNA HECTA.

GOLDEN SWIFT MOTH.

*SPECIFIC CHARACTER,*

AND

*SYNONYMS.*

Wings deflexed. Yellow brown. Two oblique, whitifh, or yellow bands, confifting of interrupted and irregular fpots, acrofs the anterior wings.

PHALÆNA HECTA: lutea, alis deflexis: primoribus fafciis duabus
    albidis obliquis punctatæ interruptis. *Fn. Sv.* 1148.——
    *Gmel. Linn. Syft. Nat. Ent. p.* 2617. *fp.* 85.
HEPIALUS HECTUS. *Fab. Ent. Syft. T.* 3. *p.* 2. *p.* 6. *Sp.* 4.
    *Degeer. Inf.* 1. *tab.* 7. *fig.* 11.

This fpecies is common in the fkirts of woods in May and June. The colours in the male Infect are more vivid than the female, and the fpots on the anterior wings in particular are of fuch a beautiful yellow, that Englifh collectors have termed this kind the Golden Swift Moth.

It commences its flight earlier in the evening than any other of the nocturnal lepidopterous infects. Its manner of flying is very fingular, and attracted the notice of Linnæus, who aptly compares it to the motion of the pendulum of a clock.

The larva is unknown: it is fuppofed to feed on the roots of plants under ground.

PLATE

# PLATE CCLXXV.

## PHALÆNA ALNIARIA.

### GENERIC CHARACTER.

Antennæ taper from the bafe. Wings in general deflexed when at reft. Fly by night.

### SPECIFIC CHARACTER

AND

### SYNONYMS.

Antennæ feathered. Wings yellow, fpeckled with brown, and marked with two tranfverfe ftreaks. Margins deeply indentated.

PHALÆNA ALNIARIA: pectinicornis alis erofis flavis fufco pul-
verulentis: ftrigis duabus fufcis. *Linn. Syft. Nat.* 2.
860. 205.—*Fn. Sv.* 1230. *Fab. Ent. Syft. T.* 3. *p.* 2.
*p.* 136. *Sp.* 24. *Schæff. Icon. tab.* 135. *fig.* 1, 2.

———————

This Infect is analogous to feveral other fpecies of Britifh Phalænæ, known amongft collectors by the indefinite term, Thorn
Moths. It is remarkable only in the larva ftate, when, from its
fimilarity in form and colour to a twig of the tree on which it feeds,
it is fuppofed to efcape the notice of its enemies. It is a dull creature, and will often remain in an oblique, or erect pofition, without
motion, or appearance of life, for feveral hours together.

It is found on fruit-trees, in the ftate of larva, in May: the Moth
appears in Auguft or September.

G PLATE

# PLATE CCLXXVI.

## CHRYSOMELA TENEBRICOSA.

### GENERIC CHARACTER.

Antennæ compofed of globular articulations, which encreafe in bulk towards the end. Thorax and elytra without margin.

### SPECIFIC CHARACTER

#### AND

### SYNONYMS.

No wings. Somewhat oval. Thorax lunated. Black, gloffed with blue, or purple.

CHRYSOMELA TENEBRICOSA: aptera ovata, atra antennis pedibufque violaceis. *Fab. Ent. Syft.* 1. 308. 3.
TENEBRIO LÆVIGATUS, apterus niger lævis, elytris lævibus thorace
lunato, fubtus cœruleis. *Linn. Syft. Nat.* 678. 29.
Chryfomela Tenebrioides. *Gmel. Linn. Syft. Nat.* 1667. 1.

---

This Creature is entirely of a black colour, gloffed, in fome fpecimens, with fhining blue; in others, with purple. Thofe colours
are particularly vivid on the legs and underfide of the Infect. It is
very common during moft part of the fummer.

I

PLATE

2

1

# PLATE CCLXXVII.

## FIG. I.

### PHRYGANEA VARIA?

NEUROPTERA.

*GENERIC CHARACTER.*

Mouth furniſhed with four palpi.  Stemmata three.  Antennæ longer than the thorax.  Firſt wings incumbent.  Second wings folded.

*SPECIFIC CHARACTER.*

Wings greyiſh brown, variegated with black, and ſpotted in the middle with white.

PHRYGANEA VARIA: alis cinereo, nigroque variis: punĉto medio diſtinĉto niveo, antennis nigris.  *Fab. Ent. Syſt. T. 2. p. 77, 103.*

---

The *Phryganea Varia* of Fabricius has not been figured by any author; but is, we ſuſpeĉt, no other than a ſmall variety of P. Grandis, and which Linnæus deſcribes as " *alis cinereo teſtaceis,* " *lineolis duabus longitudinalibus nigris, punĉto albo.*"  Fabricius has altered this definition to " *alis teſtaceis cinereo maculatis,*" in the *Entomologia Syſtematica*, and erroneouſly attributes the latter to Linnæus.

We cannot avoid, in ſupport of our opinion, noticing the ſtriking diſſimilarity between the ſeveral figures of Phryganea Grandis, in our copies of the works quoted by Linnæus and Fabricius.  *Roeſel* has two figures, one of a bluiſh grey, the other

H                                                    yellowiſh

yellowiſh brown; both vary in the markings, and are nearly twice the ſize of our ſpecimen. In *Sulzer*, the wings are ſhorter and more pointed than in the preceding work; and are rather pellucid and tinged with yellow: the markings totally different from thoſe in *Roeſel*. Thoſe in *Schæffer* are moſt like one of our ſpecimens in form and reddiſh colour, but ſtill with markings as different from it as from either of the preceding figures. We may hence conclude the marks are inconſtant, and the colours extremely variable; and that in point of ſize, the preſent ſpecies may be the *Varia* of Fabricius, ſince he notices its affinity to the P. Grandis; and adds, it is rather ſmaller.—" Nimis Affinis P. Grandis at paullo minor." *Fab. Ent. Syſt.*

## F I G. II.

### HEMEROBIUS PERLA.

#### Golden Eye.

### GENERIC CHARACTER.

Mouth armed with two teeth and four palpi. Wings deflected. Antennæ ſetaceous, and longer than the thorax.

### SPECIFIC CHARACTER.

Yellowiſh green. Wings tranſparent. Eyes golden.

Hemerobius Perla: luteo viridis alis hyalinis: vaſis viridibus. *Linn. Syſt. Nat. 2. 911. 2. Fn. Sv. 1504.*

Extremely common in ſummer. It is a very delicate little creature, greeniſh, the wings reticulated, perfectly tranſparent, with various beautiful, pearly, reddiſh, greeniſh, and yellowiſh gloſſes. The eye is large, globular, and has the appearance of gold, whence its Engliſh trivial name. Stink much before a thunder-ſtorm.

PLATE

# 278

# PLATE CCLXXVIII.

## PAPILIO POLYCHLOROS.

### WOOD TORTOISESHELL BUTTERFLY.

### *GENERIC CHARACTER.*

Antennæ clubbed at the ends. Wings erect when at reft. Fly by day.

### *SPECIFIC CHARACTER*

### AND

### *SYNONYMS.*

Wings angulated, brownifh orange, fpotted with black.

PAPILIO POLYCHLOROS: alis angulatis fulvis nigro maculatis: anticis fupra punctis quatuor nigris. *Linn. Syft. Nat.* 2. 777. 166.—*Fn. Sv.* 1057.—*Fab. Syft. Ent. t.* 3. *p.* 1. 121. *fp.*372.
*Merian Europ. tab.* 1.
*Roef. Inf.* 1. *pap.* 1. *tab.* 2.
*Schæff. Icon. tab.* 146. *fig.* 1, 2.
*Ammir. tab.* 15.
*Frifch. Inf.* 6. *tab.* 3.
*Wilks. pap.* 56. *tab.* 3. *a.* 5.

There is a ftriking fimilarity in form and colour between this fpecies and Papilio Urticæ, already figured in this work. Collectors have hence denominated the two fpecies the Tortoifefhell

H 2

Butterflies;

Butterflies; they differ, however, materially in fize, as well as manner of life. The fmalleft kind is very common, and its larva feeds on the Nettle; the prefent fpecies is found on the Elm, frequents woods, and is very fcarce.

The larva is found in June; it becomes a pupa in the fame month; and after remaining in that ftate twenty-one days, the winged Infect is produced.

PLATE

# PLATE CCLXXIX.

## PAPILIO MEGAERA.

### LEPIDOPTERA.

### *GENERIC CHARACTER.*

Antennæ clubbed at the end. Wings in general deflexed when at reft. Fly by day.

### *SPECIFIC CHARACTER*

### AND

### *SYNONYMS.*

Wings dentated yellowifh brown, with dark marks acrofs. One eye on the anterior pair: Five on the pofterior pair above, and fix beneath.

PAPILIO MEGAERA: alis dentatis luteis fufco fafciatis: anticis ocello, pofticis fupraquinis, fubtus fex. *Linn. Syft. Nat.* 2. 771. 142.—*Fab. Ent. Syft. T.* 3. *p.* 1. *p.* 94. 292.—*Schæff. Icon. tab.* 148. *fig.* 3, 4.

———————

This Butterfly is common in meadows in July. The larva is green, hairy, with a bifid tail; it feeds on grafs, and changes to the pupa ftate in June.

H 3        PLATE

# PLATE CCLXXX.

## FIG. I.

### PAPILIO NAPI.

#### GREEN-VEINED WHITE BUTTERFLY.

### GENERIC CHARACTER.

Antennæ clubbed at the end. Wings erect when-at-reft. Fly by day.

### SPECIFIC CHARACTER

#### AND

### SYNONYMS.

Wings entire, white. Beneath, veined with green.

PAPILIO NAPI: alis integerrimis albis: fubtus venis dilatis viref-
  centibus. *Linn. Syft. Nat.* 2. 760. 77.—*Fn. Sv.* 1037.
  —*Fab. Ent. Syft. T.* 3. *p.* 1. 187. 576.
  *Geoff. Inf.* 2. 70. 42.
  *Merian Europ.* 2. *tab.* 39.
  *Albin Inf. tab.* 52. *fig. F. G.*

———

Frequent in gardens in May. The larva feeds on the cabbage.

H 4

FIG.

# FIG. II.

## PAPILIO SINAPIS.

WOOD LADY, or WOOD WHITE BUTTERFLY.

*SPECIFIC CHARACTER,*

AND

*SYNONYMS.*

Wings rounded, entire, white.   Apex brown.

PAPILIO NAPI: alis rotundatis integerrimis albis: apicibus fufcis.
*Linn. Syft. Nat.* 2. 760. 79.—*Fn. Sv.* 1038.—*Fab.
Ent. Syft. T.* 3. *p.* 1. *p.* 187. *fp.* 577.
*Schæff. Icon. tab.* 97. *fig.* 8, 9, 10, 11.
*Degeer. Inf.* 2. 183. 4. *tab.* 1. *fig.* 1.
*Raj. Inf.* 116. 8.

Found in woods in May ; a fecond brood appears in Auguft.

PLATE

# 281

# PLATE CCLXXXI.

## PHALÆNA FAGANA.

### COMMON SILVER-LINE MOTH.

### *GENERIC CHARACTER.*

Antennæ, taper from the bafe. Wings in general deflexed, when at reft. Fly by night.

### *SPECIFIC CHARACTER*

#### AND

### *SYNONYMS.*

Anterior wings green, with three oblique white, or filvery ftripes acrofs each. Antennæ and feet fulvous.

PYRALIS FAGANA: alis viridibus: ftrigis tribus obliquis albis, antennis pedibufque fulvis. *Fab. Ent. Syft.* 3. *p.* 2. 243. 5.

Phalæna Fagana. *Wien. Verz.* 125. 2.
     *Roef. Inf.* 4. *tab.* 22.
     *Petiv. Gazoph. tab.* 7. *fig.* 11.
     *Ammiral Inf. tab.* 2. *fig.* 1.
     *Alb. Inf. tab.* 31.
     *Wilks. pap.* 1. *tab.* 1. *a.* 13.

---

The Phalæna Fagana is not one of the rarer Britifh fpecies; it claims attention for its fingular and beautiful appearance only. The larva is found on the oak in Auguft and September : in the

beginning

beginning of October, it spins a very extraordinary kind of covering on a leaf of the oak, and becomes a pupa within. This covering somewhat resembles a tent, or rather an inverted boat, being shuttle formed, and having a keel, or longitudinal ridge along the upper part: its colour is yellowish brown; the pupa underneath is purple. The Moth appears in the winged state in May

PLATE

1

# PLATE CCLXXXII.

## BUPRESTIS PYGMEA.

### COLEOPTERA.

### *GENERIC CHARACTER.*

Antennæ fetaceous, length of the thorax. Head half retracted, or drawn within the thorax.

### *SPECIFIC CHARACTER*

### AND

### *SYNONYMS.*

Wing cafes blue. Head and thorax brown, bronzed.

BUPRESTIS PYGMEA: elytris integris cyaneis, capite thoraceque æneis nitidis. *Fab. Ent. Syft.* 1. *p.* 2. *p.* 211. *Sp.* 110.

---

The difcovery of this minute but rare fpecies of Bupreftis in this country, is due to Alexander M'Leay, Efq.——It was found in a puddle, on the road fide, near Coombe Wood, in the month of May laft. The fpecimen is reprefented in the annexed plate; the fmalleft figure denotes the natural fize.

PLATE

283

# PLATE CCLXXXIII.

## PHALÆNA MELLONELLA.

### HONEY MOTH.

#### GENERIC CHARACTER.

Antennæ taper towards the bafe. Wings in general deflexed when at reft. Fly by night.

#### TINEA.

#### SPECIFIC CHARACTER

##### AND

#### SYNONYMS.

Anterior wings grey: pofterior part purple. Scutellum black, white at the tip.

P. TINEA MELLONELLA: alis canis pofticis purpurafcentibus: ftriga alba, fcutello nigro apice candido. *Linn. Syft. Nat.* 2. 888. 375.—*Fn. Sv.* 1383.—*Fab. Ent. Syft.* 3. *p.* 2. 305. *Sp.* 79. *Reaum. Inf.* 3. *tab.* 19. *fig.* 79.

———

The larva of this fpecies is fometimes found in bee-hives; it infinuates itfelf amongft the cells of thofe Infects, and fubfifts on the honey. It remains in the pupa ftate in a long cylindrical channel or paffage it forms in the larva ftate: the winged Infect comes forth in Auguft.

PLATE

# PLATE CCLXXXIV.

## SPHINX LIGUSTRI.

### PRIVET SPHINX, or HAWK MOTH.

#### *GENERIC CHARACTER.*

Antennæ thickeft in the middle. Wings in general deflexed when at reft. Fly by night.

#### *SPECIFIC CHARACTER*

##### AND

#### *SYNONYMS.*

Wings entire. Pofterior pair red, with three black bars acrofs. Abdomen red, with black belts.

SPHINX LIGUSTRI: alis integris pofticis rufis; fafciis tribus nigris, abdomine rubro: cingulis nigris. *Linn. Syft. Nat.* 2. 799. 8. *Fn. Sv.* 1087.

Sphinx Liguftri: *Fab. Ent. Syft. T.* 3. *p.* 1. 374. 55.
   *Roef. Inf.* 3. *tab.* 5.
   *Degeer. Inf.* 1. *tab.* 1. *fig.* 6.
   *Schæff. Elen. tab.* 116. *fig.* 2.
   *Albin. Inf. tab.* 7. *fig.* 10.
   *Efp. Inf.* 2. *tab.* 6.
   *Reaum. Inf.* 2. *tab.* 20, *fig.* 1—4.

---

The larva of this beautiful fpecies is found very frequently on the Privet in the months of July and Auguft. It buries itfelf in the earth preparatory to its becoming a pupa, and comes forth in the fly ftate in June following.

PLATE

# PLATE CCLXXXV.

## CASSIDA MACULATA.

### COLEOPTERA.

### *GENERIC CHARACTER.*

Antennæ nearly filiform, but encreafing in bulk towards the extremity. Margin of the elytra broad. Head concealed under the thorax.

### *SPECIFIC CHARACTER*

#### AND

### *SYNONYMS.*

Greenifh, variegated with fpots of black on the elytra, particularly along the future of the back.

CASSIDA MACULATA: viridibus elytris rarius, futura dorfali confertius nigro maculatis. *Linn. Syft. Nat.* 2. 575. 6. *Fab. Syft. Ent.* 88. 2.
*Caffida* viridis maculis nigris variegata. *Geoff. Inf.* 1. 314. 5. *tab.* 5. *fig.* 6.

---

Has been fuppofed only a variety of Caffida viridis, but is evidently a diftinct fpecies, and is very uncommon in England.

Fig. I. The natural fize.

I                                          PLATE

## F I G. II.

## CHRYSOMELA GOETTINGENSIS.

### *SPECIFIC CHARACTER.*

Oval black, gloffed with purple. Legs purple. Tarfi reddifh.

CHRYSOMELA GOETTINGENSIS: ovata atra pedibus violaceis:
plantis rufis. *Linn. Syft. Nat.* 2. 586. 4.—*Fn. Sv.* 506.
*Fab. Ent. Syft. T.* 1. *p.* 309.
*Degeer. Inf.* 5. 298. 8.

————————————

A rare fpecies in this country; very common in Germany.

P L A T E

# PLATE CCLXXXVI.

## FIG I.

### CHRYSOMELA MARSHAMI.

*GENERIC CHARACTER.*

Antennæ articulated, larger towards the end. Thorax and elytra without margin.

*SPECIFIC CHARACTER.*

Thorax greenifh gold. Wing-cafes coppery, irregularly punctated.

CHRYSOMELA MARSHAMI: thorace viridi æneo, elytris cupreis vagè punctatis.

———————

As the Britifh Coleoptera are in few inftances remarkable for that fplendid glow and gaiety of colours which diftinguifh thofe of warmer climates, we are more inclined to admire the beauty of this recently difcovered fpecies of Chryfomela. It appears not to have been noticed by any preceding author; and as it remains with us to defignate fome fpecific appellation, we have named it *Marfhami*, in compliment to that eminent entomologift T. Marfham, Efq.—a name, perhaps, the more appropriate, as it is known among that gentleman's fcientific friends the world will foon be favoured with his invaluable papers on Britifh Coleoptera.

This Infect is nearly allied to C. Faftuofa and C. Hypericum.— Found in Norwood laft May.

FIG.

# 287

# PLATE CCLXXXVII.

## FIG. I.

## PHALÆNA PAPILIONARIA.

### LARGE EMERALD MOTH.

### *GENERIC CHARACTER.*

Antennæ taper from the bafe. Wings in general deflexed when at reft. Fly by night.

### *SPECIFIC CHARACTER.*

Antennæ feathered. Wings green: a continued ftreak of whitifh femicircular marks acrofs the middle of each, and an interrupted or half ftreak of the fame colour below it.

PHALÆNA PAPILIONARIA: pectinicornis alis fubrepandis viridibus: ftriga fefquialtera repanda. *Linn. Syft. Nat.* 2. 864. 225.—*Fn. Sv.* 1247.
*Wien. Verz.* 96. 1.
*Roef. Inf.* 4. *tab.* 18. *fig.* 3.

---

A rare and very elegant Britifh fpecies. The larva is green, with about ten incurvated fpines or hooks along the back. It is found in this ftate on the Birch and Alder in June, changes to the pupa the latter end of the fame month, and appears on the wing fourteen days after.

The pupa is green variegated with yellow.

K                                        FIG.

## F I G. II.

### PHALÆNA PENNARIA.

*SPECIFIC CHARACTER.*

Antennæ feathered ; Wings indented reddifh : two dark ftreaks acrofs the anterior pair, and a diftinct white fpot near the apex.

PHALÆNA PENNARIA: pectinicornis alis fubdentatis rufefcentibus : ftrigis duabus fufcis punctoque apicis albo.
*Linn. Syft. Nat.* 2. 861. 209.—*Fab. Ent. Syft.* 3. *p.* 2. 132. 14.

The larva fmooth reddifh brown ; found on Fruit-trees.

## F I G. III.

### PHALÆNA BILINEATA.

*SPECIFIC CHARACTER.*

Antennæ fetaceous, Wings yellow undulated with brown ftreaks, and a broad wave acrofs the anterior pair.

PHALÆNA BILINEATA : feticornis alis luteis teftaceo undatis : fafcia repanda, margine fufco. *Linn. Syft. Nat.* 2. 868. 245.—*Fn. Sv.* 1284.
*Clerk. Icon. tab.* 6. *fig.* 13.

Very abundant in White-thorn hedges during moft part of the Summer.

PLATE

# 288

1

1

2

2

# PLATE CCLXXXVIII.

## FIG. I. I.

## CICADA NITIDULA.

### GENERIC CHARACTER.

Roftrum bent inwards. Antennæ fetaceous. Wings membranaceous declining along the fides of the body.

### SPECIFIC CHARACTER.

Yellow. Wing-cafes tranfparent, whitifh, with two dark tranfverfe bars.

CICADA NITIDULA: flava, elytris hyalino albis, faciis duabus nigris. *Fab. Ent. Syft.* 4. *p.* 46. *n.* 87.

––––––––––––

This minute fpecies is reprefented in the natural fize at Fig. I. The upper figure exhibits its magnified appearance.

FIG.

## F I G. II.

## CICADA FLAVOSTRIATA.

### SPECIFIC CHARACTER.

Black. Head and Thorax tranfverfely ftreaked with yellow; ftreaks on the wing-cafes of the fame colour, difpofed longitudinally.

CICADA FLAVOSTRIATA: -nigra, capite thoraceque tranfverfe elytrifque longitudinaliter flavo-ftriatis.

---

This Infect has been confounded with the Cicada ftriata of *Linnæus, Faun. Suec.* 887. and *Syft. Nat.* 709. *n.* 30. in general; but it is certainly not that fpecies. Linnæus refers to *Geoff.* 1. *p.* 424. *n.* 20. for *C. ftriata,* wherein it is thus defcribed; " Head pale green, with two black points in front, and four near the bafe; Thorax of the fame colour, marked with feveral lefs diftinct black fpots alfo." We therefore confider our C. flavoftriata as a new fpecies.

PLATE

# LINNÆAN INDEX

### TO

## VOL VIII.

### COLEOPTERA.

L            LEPI-

# I N D E X.

## LEPIDOPTERA.

NEUROPTERA.

# INDEX.

## NEUROPTERA.

---

L 2                    ALPHABETICAL

# ALPHABETICAL INDEX

TO

# VOL. VIII.

Marſhami,

# INDEX.

THE

# NATURAL HISTORY

OF

# BRITISH INSECTS;

EXPLAINING THEM

IN THEIR SEVERAL STATES,

WITH THE PERIODS OF THEIR TRANSFORMATIONS
THEIR FOOD, OECONOMY, &c.

TOGETHER WITH THE

## HISTORY OF SUCH MINUTE INSECTS

AS REQUIRE INVESTIGATION BY THE MICROSCOPE.

THE WHOLE ILLUSTRATED BY

# COLOURED FIGURES,

DESIGNED AND EXECUTED FROM LIVING SPECIMENS.

---

By E. DONOVAN.

---

VOL. IX.

---

LONDON:

PRINTED BY BYE AND LAW, ST. JOHN'S SQUARE, CLERKENWELL,

FOR THE AUTHOR,

And for F. and C. RIVINGTON, Nº 62, ST. PAUL'S CHURCH-YARD.

MDCCC.

S

# 289

# THE

# NATURAL HISTORY

OF

# BRITISH INSECTS.

---

## PLATE CCLXXXIX.

### SPHINX ATROPOS.

### DEATH HEAD, or BEE TIGER MOTH.

#### GENERIC CHARACTER.

Antennæ thickeſt in the middle. Wings deflexed, the outer margin declining towards the ſides.

#### SPECIFIC CHARACTER

AND

#### SYNONYMS.

Wings entire; poſterior pair yellow, barred acroſs with brown. Abdomen yellow, with black rings.

SPHINX ATROPOS: alis integris: poſticis luteis; faſciis fuſcis, abdomine luteo: cingulis nigris. *Lin. Syſt. Nat. 2.* 799. 9.—*Muſ. Lud. Ulr.* 348.

A 2                                                    *Reaum.*

*Reaum. Inf.* 1. *tab.* 14.
*Roef. Inf.* 3. *tab.* 1. 1.
*Haffelquift. Itin.* 407. 104. 105.
*Schæff. Icon. tab.* 99. *fig.* 1. 2.
*Efp. Inf.* 2. *tab.* 7.
*Sulz. Inf. tab.* 15. *fig.* 88.
*Albin. Inf. tab.* 6.
*Wilks. pap.* 9. *tab.* 1. *B.* 1.

---

The Sphinx Atropos is a magnificent creature, and the largeft of the European Lepidopterous Infeĉts. The charaĉteriftic marks of this fpecies are very fingular; on the thorax in particular the figure of a human fkull is ftrongly depiĉted. Thefe Infeĉts have been deemed a prefage of fome approaching calamity, by the peafantry in countries where they have appeared by chance; and Linnæus has himfelf named it after one of the three Fates, of the Heathen Mythology.

This fpecies feems no where common. In this country it is rare. We have an Englifh Specimen in the winged ftate, and once met with its larva, of a full fize, but it died before it became a pupa.

P L A T E

# PLATE CCXC.

### THE

## LARVA

#### OF

## SPHINX ATROPOS.

### DEATH HEAD, or BEE TIGER MOTH.

———————

This Specimen was found on the Jaſmine, the latter end of Auguſt. It is ſaid to feed alſo on Potatoes and Green Elder. It appears in the Fly ſtate in July.

PLATE

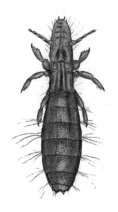

1

# PLATE CCXCI.

## PEDICULI COLUMBÆ.

### PIGEON'S LOUSE.

#### APTERA.

### GENERIC CHARACTER.

Six Feet. Eyes two. Mouth contains a fting. Antennæ length of the Thorax.

### SPECIFIC CHARACTER

#### AND

### SYNONYMS.

Body flender, thickeft towards the end, whitifh, a ferruginous line along each fide.

PEDICULUS COLUMBÆ : corpore filiformi ferrugineis poftice clavato. *Fab. Syft. Ent.* 809. 31.
*Pediculus* oblongus filiformis albicans, corporis lateribus utrinque ferrugineis. *Geoff. Inf.* 2. 599. 7.
Pulex Columbæ Majoris. *Redi de Infectis tab.* 2. *fig.* 1.

———

This is the kind of Loufe found on the common Pigeon. It differs fpecifically from thofe of other Birds and Animals, as may be conjectured from its fignificant name, *P. Columbæ.*

It

It is fuppofed that almoft every creature has its peculiar fpecies of Pediculi, but not more than fifty diftinct fpecies have been hitherto afcertained; and nearly the whole of thefe belong to the feathered tribe. *Redi, Linnæus, Fabricius,* and other Authors diftinguifh many of thefe Infects by the names of thofe creatures they infeft, as *P. Vulturis, Orioli, Cuculi, Cygni, Pavonis, Phafiani,* &c. &c.

Fig. I. Natural Size.    Fig. II. Magnified.

P L A T E

# 292

# PLATE CCXCII.

## PAPILIO CARDUI.

PAINTED LADY BUTTERFLY.

### GENERIC CHARACTER.

Antennæ clubbed.   Wings erect when at reft.   Fly by day.

### SPECIFIC CHARACTER

AND

### SYNONYMS.

Wings indented, above orange variegated with black and white ; beneath, four eyes on the pofterior pair.

PAPILIO CARDUI: alis dentatis fulvis albo nigroque variegatis : pofticis fubtus ocellis quatuor. *Linn. Syft. Nat. 2.*
*774. 157.—Fn. Sv.* 1054
*—Fab. Ent. Syft. T. 3. p.* 1. 104. *fp.* 320.
*Schæff. Icon. tab.* 97. *fig.* 5, 6.
*Ernft. Pap. Europ.* 1. *tab.* 7.
*Albin. Inf. tab.* 56.
*Cram. Inf.* 3. *tab.* 26. *fig. E. F.*
*Reaum. Inf.* 1. *tab.* 26. *fig.* 11, 12.
*Hoeffn. Inf. tab.* 7. *fig.* 3.

The Painted Lady Butterfly is a local fpecies, and therefore not very common.   In fome feafons, thefe Infects appear in confiderable numbers, and then again are not feen for feveral years.   They

B                                                                were

were taken in abundance in the fummer of 1795, in many parts of the kingdom, and particularly in Manchefter; but fince that time, few, if any, have been met with.

In point of beauty, this fpecies has an immediate claim to the notice of Englifh Entomologifts. Its larva feeds on nettles, thiftles, docks, and other herbage by the fides of ditches, and changes to the pupa ftate about the middle or latter end of July; the winged Infect appears twelve days after.

P L A T E

# PLATE CCXCIII.

## FIG. I.

### PHALÆNA ULMATA.

SCARCE ELM MOTH.

#### GENERIC CHARACTER.

Antennæ taper from the bafe. Wings in general deflexed when at reft. Fly by night.

#### SPECIFIC CHARACTER

AND

#### SYNONYMS.

Antennæ fetaceous. Wings white, with a double row of pale black fpots acrofs the middle: a ferruginous brown fpot at the bafe, and another at the pofterior margin of the firft pair: a fimilar fpot in the interior margin of the fecond pair alfo.

PHALÆNA ULMATA: feticornis alis albidis: fafciis duabus ferrugineo fufcis; poftica maculari. *Fab. Ent. Syft.* 3.
*p.* 2. *p.* 176. *Sp.* 171.
Phalæna pantaria pectinicornis alis albis: fafcia maculari flavicante,
abdomine luteo nigro punctato. *Linn. Syft. Nat.* 2.
863. 218.

This fpecies bears fome affinity to the Phalena Groffulariata, or Currant Moth. It is very rare, and has been hitherto found only in Yorkfhire. It appears the third week in June. The larva feeds on the elm: it is green, ftreaked with black, and has a black head; the pupa blueifh.

B 2

FIG.

### F I G. II.

## PHALÆNA MARGINATA.

*SPECIFIC CHARACTER*

AND

*SYNONYMS.*

Antennæ fetaceous. Wings white, with a deep irregular brown margin of interrupted fpots.

PHALÆNA MARGINATA: feticornis alis omnibus albis: margine
    exteriori limbo fufco interrupto. *Linn. Syft. Nat.* 2.
    870. 257.—*Fn. Sv.* 1279.
    *Sulz. Inf. tab.* 16. *fig.* 96.
    *Geoff. Inf.* 2. 139. 60.
    *Clerk. Phal. tab.* 2. *fig.* 5.

---

Phalæna *Geometra* Marginata\* is rather common. It lives on the nut, and is found in the winged ftate in May.

---

  \* As the fpecific name *Marginata* occurs in the defcription of a Moth in Plate CL. it may be proper to obferve, that the Infect there figured, is *Noctua Marginata* of Fabricius.—The *fectional* divifions of the Phalænæ muft be particularly attended to, when the fame fpecific name occurs more than once.

F I G.

PLATE CCXCIII. 13

## FIG. III.

### PHALÆNA·PRUNARIA, *Var.*

FEMALE ORANGE MOTH.

*Male,* Plate 23. *Br. Inf.*

---

The male of Phalæna Prunaria is reprefented and defcribed in the early part of our Britifh Infects; but it differs fo confiderably from the fingular variety of the fame fpecies now before us, that we prefume to introduce it in the annexed plate of *Geometræ.*

B 3                    PLATE

 *1*

# PLATE CCXCIV.

## MUSCA SOLSTITIALIS.

### *GENERIC CHARACTER.*

The mouth formed by a foft flefhy probofcis, with two lateral lips. No palpi.

### *SPECIFIC CHARACTER*

AND

### *SYNONYMS.*

Antennæ furnifhed with a lateral hair. Wings white, with four fomewhat connected black bars acrofs. Scutellum yellow.

Musca Solstitialis: antennis fetariis, alis albis: fafciis quatuor connexis nigris, fcutello flavo. *Linn. Syft. Nat.* 2. 999. 127.—*Fn. Sv.* 1879.
*Degeer. Inf.* 6. 42. 16. *tab.* 2. *fig.* 10, 11.
La mouche des tetes de Chardons. *Geoff. Inf.* 2. 499. 14.

———

Found in the middle of fummer on thiftles, and is an elegant object for the microfcope.

B 4 . PLATE

# PLATE CCXCV.

## PAPILIO COMMA.

### PEARL SKIPPER BUTTERFLY.

### *GENERIC CHARACTER.*

Antennæ clubbed at the ends. Wings erect when at rest. Fly by day.

*Plebeii Urbicolæ.*

### *SPECIFIC CHARACTER.*

Wings entire, divaricated, brown, having a black ſtreak along the middle of the anterior pair. Beneath, ſpotted with white.

PAPILIO COMMA : alis integerrimis divaricatis fulvis: punctis albis
    lineolaque nigra. *Linn. Syſt. Nat. p.* 793. 256.
HESPERIA COMMA : *Fab. Ent. Syſt. t.* 3. *p.* 1. *p.* 325. 233.
PAPILIO COMMA : *Wien. Verz.* 159. 4.

———————

In the beginning of Auguſt, 1772, a brood of theſe Inſects were taken near Lewes in Suſſex, by the late Mr. Green; and we believe no other ſpecimens have been taken ſince that period.

It is not very unlike the Papilio Sylvanus of Fabricius, but may be readily diſtinguiſhed from it by the ſquare ſpots on the under-ſide being perfectly white.

PLATE

# PLATE CCXCVI.

## SPHINX PINASTRI.

### PINE HAWK MOTH.

### *GENERIC CHARACTER.*

Antennæ thickeſt in the middle. Wings deflexed when at reſt. Fly ſlow, morning and evening.

### *SPECIFIC CHARACTER*

AND

### *SYNONYMS.*

Wings entire. Greyiſh white: three contiguous black lines in the middle of the anterior pair. Abdomen browniſh, with bands of white.

SPHINX PINASTRI: alis integris canis: anticis lineolis tribus con-
fertis nigris, abdomine fuſco: cingulis albis. *Linn.*
*Syſt. Nat.* 2. 802. 22.—*Fn. Sv.* 1088.
*Fab. Ent. Syſt. T.* 3. *p.* 1. *p.* 367. 35.
*Eſp. Inſ.* 2. *tab.* 12.
*Roeſ. Inſ.* 1. *phal.* 1. *tab.* 6.
*Reaum. Inſ.* 1. *tab.* 13. *fig.* 8.

---

We have only a traditionary report that Sphinx Pinaſtri has been ſometimes found in Scotland; but as it is generally admitted, on that authority, to a place in the cabinets of Engliſh Inſects, we cannot refrain inſerting it in the preſent work.

It is an European Inſect, and in particular is found in the Pine foreſts of Germany. Roeſel has figured it with the larva and pupa,

in

in the plate above quoted ; and as we may, perhaps, never meet with it in that ftate, we conceived the copies of them in the annexed plate, would at leaft be fatisfactory to fuch fubfcribers as have not the works of that German author.

**P L A T E**

1

# PLATE CCXCVII.

## CIMEX BICOLOR.

### BLACK AND WHITE FIELD BUG.

### GENERIC CHARACTER.

Roſtrum inflected. Antennæ longer than the thorax. One wing-caſe folded over the other. Feet formed for running

### SPECIFIC CHARA ᵀER

### AND

### SYNONYMS.

Black. Wing-caſes variegated with black and white; a ſemi-circular white ſpace in the middle. Wings tranſparent and whitiſh.

CINEX BICOLOR: niger elytris albo variis, alis albis. *Linn. Syſt.*
    *Nat.* 2. 722. 55.—*Fn. Sv.* 936.—*Fab. Ent. Syſt. T.* 4.
    *p.* 121. *ſp.* 161.
La Punaiſe noire à quatre taches blanches. *Geoff. Inſ.* 1. *p.* 470. *ſp.* 73.
    *Schæff. Icon. tab.* 41. *fig.* 8. 9.
    *Stoll. Cimic. tab.* 32. *fig.* 224.
    *Petiv. Gazoph. tab.* 14. *fig.* 7.

The natural ſize is repreſented at Fig. I.

PLATE

# PLATE CCXCVIII.

## PHALÆNA TRIPLACIA.

### SPECTACLE MOTH.

### *GENERIC CHARACTER.*

Antennæ taper from the bafe. Wings in general deflexed when at reft. Fly by night.

\*\* *Noctua.*

### *SPECIFIC CHARACTER*

### AND

### *SYNONYMS.*

Thorax crefted. Wings deflexed; firft pair greyifh, with a double ferruginous arch; at the bafe, and another in an oppofite direction near the apex.

PHALÆNA TRIPLACIA: critata alis deflexis; anticis arcu duplici contrario maculifque tribus glaucis intermediis. *Linn. Syft. Nat.* 2. 854. 175.—*Fn. Sv.* 1202. *Fab. Ent. Syft. T.* 3. *p.* 2. 117. 354. *Degeer. Inf.* 1. *tab.* 6. *fig.* 20. 21. *Merian. Europ. tab.* 97.

This Infect is remarkable for a fanciful kind of marking that encircles the eyes, and feems to refemble a pair of fpectacles. It is found in the winged ftate the fourth week in June.

PLATE

# PLATE CCXCIX.

## PHALÆNA ROBORIS.

### *GENERIC CHARACTER.*

Antennæ taper from the bafe. Wings in general deflexed when at reft. Fly by night.

\* \* *Noctua.*

### *SPECIFIC CHARACTER*

#### AND

### *SYNONYMS.*

Wings deflexed, greyifh: two undulated white waves acrofs the anterior pair: a white fpace in the middle, having a lunar black mark in its center.

PHALÆNA ROBORIS: lævis alis deflexis cinereis: ftrigis duabus undatis albis, macula centrali nivea: lunula nigra. *Fab. Ent. Syft.* 3. *p.* 2. *p.* 35. *fp.* 90.

———————————————

A fcarce fpecies; it is found on the nut-tree.

C PLATE

# PLATE CCC.

## TIPULA POMONÆ.

### GENERIC CHARACTER.

Head lengthened out. Upper jaw arched. Palpi two, curved, longer than the head. Probofcis fhort, and bent inwards.

### SPECIFIC CHARACTER

#### AND

### SYNONYMS.

Shining black. Wings whitifh, with a central dark fpot in the anterior margin. Thighs ferruginous.

TIPULA POMONÆ: glabra nigra alis lacteis: puncto nigro, femoribus ferrugineis. *Fab. Ent. Syft. T. 4. p.* 249. *fp.* 74.

---

We met with both fexes of this uncommon Infect in Coombe Wood, Surry, in the beginning of June, 1798.—The fpecimen Fabricius defcribed was taken in this country alfo, May 13, on a fruit-tree.

PLATE

# PLATE CCCI.

## CICINDELA RIPARIA.

### GENERIC CHARACTER.

Antennæ fetaceous. Jaws advanced and armed with teeth. Eyes prominent. Thorax roundifh and margined.

### SPECIFIC CHARACTER

#### AND

### SYNONYMS.

Greenifh, bronzed, with many excavated round fpots on the wing-cafes.

CICINDELA RIPARIA : viridi-ænea, elytris punǎis latis excavatis.
   *Fn. Suec.* 741. *Gmel. Linn. Syft. T.* 1. *p.* 4. *p.* 1925.
   *fp.* 10.
Cicindela viridi-ænea maculis rotundis excavatis grifeo-viridibus.
   *Degeer. Inf.* 4. *p.* 117. *n.* 4. *t.* 4. *f.* 9.

------------

We found this beautiful fpecies in fome plenty in a little marfhy fpot behind the town of Newton, on the fea fhore of Glamorgan-fhire. Gmelin fays it is found in wet places, and obferves that its colour often varies.

It is a minute infeǎ, and is reprefented magnified in the annexed plate.

D

PLATE

# PLATE CCCII.

## PAPILIO AGLAJA.

### SILVER SPOT FRITTILARY BUTTERFLY.

#### GENERIC CHARACTER.

Antennæ clubbed at the ends. Wings erect when at reft. Fly by day.

#### SPECIFIC CHARACTER

AND

#### SYNONYMS.

Wings dentated, fulvous, with black fpots. Twenty filver fpots on the under fide of the pofterior wings.

PAPILIO AGLAJA: alis dentatis fulvis nigro maculatis: fubtus 21. maculis argenteis. *Linn. Syft. Nat.* 2. 785.
211.—*Fn. Sv.* 1064.—*Fab. Ent. Syft. T.* 3. *p.* 1.
*p.* 144. *fp.* 442.
*Wilks Pap. tab.* 2. *a.* 12.
*Efp. Pap.* 1. *tab.* 17. *fig.* 3.
*Schæff. Icon. tab.* 7. *fig.* 1, 2.

---

A very beautiful and not uncommon Britifh fpecies ; the Larva feeds on the Violet, &c. ; it is of a dirty black colour, fpotted with brown, and armed with long fpines, as in P. Antiopa. This Larva is found in May, changes to the pupa ftate the latter end of the fame month, and appears twenty-one days after a winged infect.

D 2                                   PLATE

# PLATE CCCIII.

## DYTISCUS 2 PUNCTATUS.

### Two Spot Boat Beetle.

### GENERIC CHARACTER.

Antennæ either setaceous, or furnished at the end with a perfoliated capitulum. Hind feet formed for swimming, and hairy.

### SPECIFIC CHARACTER

#### AND

### SYNONYMS.

Black brown. Thorax yellow with two black points: wing-cases variegated with yellow and brown.

DYTISCUS 2 PUNCTATUS: ater thorace flavo: punctis duobus nigris, elytris flavo fuscoque variis. *Fab. Ent. Syst.* *T. 3. p. 1. p. 192. sp. 22.*

———————————

Fabricius describes this as a German insect. It has not been figured by any author, and is uncommon in Great Britain. Lives in the water.

D 3        PLATE

The larva is beautifully variegated with red, and tender fhades of green and yellow on a whitifh ground: it feeds on the wormwood, and becomes a pupa within a cafe, or fpinning. Found in the winged ftate in July.

PLATE

# PLATE CCCV.

## CERAMBYX OCULATUS.

\*\* *Saperda.*

### GENERIC CHARCTER.

Antennæ articulated, tapering towards the ends. Thorax either armed with fpines, or gibbous. Wing-cafes throughout of equal breadth.

### SPECIFIC CHARACTER

AND

### SYNONYMS.

Cylindrical: Thorax without fpines, yellow, with two black fpots. Wing-cafes grey with linear ftreaks of excavated black points.

CERAMBYX OCULATUS: thorace mutico cylindrico luteo: punctis duobus nigris, elytris faftigiatis linearibus nigris. SAPERDA. *Linn. Faun. Suec.* 664.—Cerambyx ferrugineo-rufus, elytris nigro cinereis punctis excavatis nigris. *Uddm. Diff.* 31—*Gmel. Linn. Syft. T.* 1. *p.* 4. *p.* 1841. *fp.* 60.

SAPERDA OCULATA. *Fab. Ent. Syft. T.* 1. *p.* 2. *p.* 308. *Schæff. Icon. tab.* 128. *fig.* 4.

The

This infect is defcribed and figured, by fome of the continental
writers on entomology, as a native of France, Italy, and Germany,
but has not, we believe, been hitherto noticed as a Britifh fpecies.
Like other local infects it is faid to be extremely common in the
Ifle of Ely, Cambridgefhire, and perhaps is not found in any other
part of the country.

P L A T E

# PLATE CCCVI.

## PHALÆNA QUADRA.

### SPOTTED FOOTMAN MOTH.

### GENERIC CHARACTER.

Antennæ taper from the bafe. Wings in general deflexed when at reft. Fly by night.

### SPECIFIC CHARACTER

#### AND

### SYNONYMS.

Thorax fmooth. Wings depreffed, yellow, with two dark blue fpots on the anterior pair.

PHALÆNA QUADRA: lævis alis depreffis luteis: anticis punctis duobus cyaneis. *Linn Syft. Nat.* 2. 840. 14. —*Fab. Ent. Syft. T.* 3. *p.* 2. *p.* 24. *fp.* 54. *Schæff. Elem. tab.* 98. *fig.* 5. *Roef. Inf.* 1. *phal.* 2. *tab.* 17.

---

The larva of this fpecies rarely occurs, and the winged Infect is not common. Linnæus has taken his specific character of this Moth from the four blue fpots on the anterior wings : it is therefore neceffary to obferve, that the other fex has no fuch fpots, and has erroneoufly been made a diftinct fpecies by the fame author, becaufe it was deftitute of them.—Found in the winged ftate in May and June.

PLATE

# PLATE CCCVII.

## PHALÆNA POPULI.

### DECEMBER MOTH.

### GENERIC CHARACTER.

Antennæ taper from the bafe.    Wings in general deflexed when at reft.    Fly by night.

### SPECIFIC CHARACTER

#### AND

### SYNONYMS.

Brown : an irregular pale ftreak acrofs the anterior pair, and a fmaller one near the bafe.    A fingle pale ftreak acrofs the pofterior pair.

PHALÆNA POPULI : fufca antice pallida, alis reverfis fufcefcen-
  tibus : ftriga fefquialtera repanda albida. *Linn. Syft.*
  *Nat.* 2. 818. 34.—*Fn. Sv.* 1101.
  —*Fab. Ent. Syft. T.* 3. *p.* 2. *p.* 429. *fp.* 70.
  *Wien. Verz.* 58. 9.
  *Roef. Inf.* 1. *phal.* 2. *tab.* 60.

―――――

We feldom meet with this interefting fpecies, for it is found both in the larva and perfect ftate in the feafon, when few collectors are difpofed to feek for it.    It feeds on the white-thorn, becomes a pupa in November, and the Moth appears in December as its trivial Englifh name implies.

PLATE

# PLATE CCCVIII.

## STAPHYLINUS ERYTHROPTERUS.

### *GENERIC CHARACTER.*

Antennæ moniliform. Elytra not half the length of the abdomen. Wings folded, and concealed under the elytra.

### *SPECIFIC CHARACTER*

**AND**

### *SYNONYMS.*

Black. Wing-cafes, antennæ, and legs red.

STAPHYLINUS ERYTHROPTERUS: ater, elytris, antennarum bafi
     pedibufque rufis.—*Fn. Suec.* 842. *Gmel. Linn. T.* 1.
     *p.* 4. *p.* 2027. *fp.* 4.
     *Fab. Ent. Syft.*
     *Degeer. Inf.* 4. *p.* 21. *n.* 6.
     *Schæff. Elem. tab.* 117.
     —*Icon. tab.* 2. *fig.* 2.

———————————————

Found in general in moift or fandy places.

PLATE

# PLATE CCCIX.

## PHALÆNA CORYLI.

### NUT-TREE TUSSOCK MOTH.

#### *GENERIC CHARACTER.*

Antennæ taper from the bafe.   Wings in general deflexed when at reft.   Fly by night.

#### *SPECIFIC CHARACTER.*

Wings deflexed greyifh : a broad ferruginous fpace acrofs the anterior wings, marked in the middle with two black points encircled with white.

PHALÆNA CORYLI: alis deflexis glaucis: fafcia ferruginea ;
    puncto nigro albo annulato, thorace variegato.
    *Linn. Syft. Nat. 2. 823. 50.—Fn. Sv. 1123.—Fab.*
    *Ent. Syft. T. 3. p. 2. p. 444. fp. 114.*
    *Degeer Inf. 1. tab. 18. fig. 4. 5.*
    *Roef. Inf. 1. phal. 2. tab. 58.*
    *Albin. Inf. tab. 90.*

Found on the nut-tree in Coombe Wood in the larva ftate in May : and formed a fine web within the leaves, where it became a pupa. The Moth appeared in July.

E        PLATE

# PLATE CCCX.

## PHALÆNA VERNARIA.

### GREEN HOUSE-WIFE MOTH.

#### *GENERIC CHARACTER.*

Antennæ taper from the base. Wings in general deflexed when at reft. Fly by night.

#### *SPECIFIC CHARACTER.*

Antennæ feathered ; fetaceous at the apex. Wings angulated, green, with two equidiftant whitifh bars acrofs : margin of alternate fpots of brown and white.

PHALÆNA VERNARIA: pectinicornis alis angulatis virefcentibus: ftrigis duabus albis repandis, antennis apice fetaceis. *Linn. Syft. Nat.* 2. 858. 195.—*Fn. Sv.* 1227. *Fab. Ent. Syft. T.* 3. *p.* 2. *p.* 129. 169. *fp.* 3.

---

A very common and pretty little fpecies. Is found on the jaf-mine and honeyfuckle.

PLATE

# PLATE CCCXI.

## PHALÆNA PRONUBA.

### YELLOW UNDERWING MOTH.

### *GENERIC CHARACTER.*

Antennæ taper from the bafe. Wings in general deflexed when at reft. Fly by night.

### *SPECIFIC CHARACTER*

### AND

### *SYNONYMS.*

Thorax crefted. Wings incumbent. Firft pair variegated brown and grey. Second pair yellow, with a black band near the margin.

PHALÆNA PRONUBA: criftata, alis incumbentibus, pofticis rubris,
        fafcia atra fubmarginali. *Fab. Syft. Ent.* 603. 55.
        —*Sp. Inf.* 2. *p.* 222. 73.
Phalæna pronuba. *Linn. Syft. Nat.* 2. 842. 121.—*Fn. Sv.* 1167.
Phalæna antennis fetaceis, alis brunneis aut cinereis, pofticis luteis,
        fafcia maginali nigra. *Degeer. Inf. Verf. Germ.* 2.
        1. 288. 1.
        *Goed. Inf.* 1. *tab.* 14.
        *Frifch. Inf.* 4. *tab.* 32.
        *Ammiral. tab.* 8.
        *Schæff. Icon. tab.* 196. *fig.* 1. 2.
        *Geoffr. Inf.* 2. 146. 76.

F

The

The larva of this beautiful, though common Moth, is found in the month of May, feeding on the roots of grafs, &c. &c. ; changes to the pupa, and appears in the winged ftate in Auguft.

PLATE

# PLATE CCCXII.

## PAPILIO EUPHROSYNE.

### PEARL BORDER FRITILLARY BUTTERFLY.

### LEPIDOPTERA.

### *GENERIC CHARACTER.*

Antennæ clubbed at the end. Wings erect when at reft. Fly by day.

### *SPECIFIC CHARACTER*

#### AND

### *SYNONYMS.*

Wings indented. Upper fide fulvous brown with black fpots. A border of filver fpots on the underfide.

PAPILIO EUPHROSYNE: alis dentatis fulvis nigro maculatis: fubtus maculis novem argenteis. *Linn. Syft. Nat.* 2. 786, 214. *Fn. Sv.* 1069.
*Fab. Ent. Syft. T. p.* 1. *p.* 147. *fp.* 450.
*Geoffr. Inf.* 2. 44. 1**.
*Degeer. Inf.* 2. *tab.* 1. *fig.* 10. 11.
*Efp. pap.* 1. *tab.* 18. *fig.* 3.

———————————

An elegant fpecies. Is common in woods, and appears in the winged ftate in May.

F 2                              PLATE

# PLATE CCCXIII.

## CARABUS NITENS.

SHINING CARABUS.

### GENERIC CHARACTER.

Antennæ fetaceous. Thorax heart-fhaped truncated at the apex. Elytra margined.

### SPECIFIC CHARACTER.

AND

### SYNONYMS.

No wings. Elytra rugged, with feveral longitudinal ridges, green margin reddifh gold. Legs black.

CARABUS NITENS: apterus elytris porcatus fcabris viridibus: margine aureo, pedibus nigris. *Linn. Syft. Nat.* 2. 669. 6.—*Fn. Sv.* 185.

Carabus nitens. *Eab. Ent. Syft. T. I. p. 131. fp.* 30

Carabus nitens. *Paykull. Monogr.* 24. 12.

Carabus aureus. *Degeer. Inf.* 4. 94. 9.

*Schæff. Icon. tab.* 51. *fig.* 1.

*Sulz. Hift. Inf. tab.* 7. *fig.* 3.

The Carabus nitens is a very rare and recently difcovered fpecies in Great Britain. It is lefs uncommon in other parts of Europe, and efpecially in Germany, from whence the Englifh collectors are ufually furnifhed with fpecimens for their cabinets.

The fmalleft figure denotes the natural fize.

F 3

PLATE

# PLATE CCCXIV.

### SPHINX PORCELLUS.

### SMALL ELEPHANT HAWK MOTH.

### LEPIDOPTERA.

### *GENERIC CHARACTER.*

Antennæ thickeſt in the middle. Wings deflected when at reſt. Fly ſlow, morning and evening.

### *SPECIFIC CHARACTER*

#### AND

#### *SYNONYMS.*

Wings entire, variegated with yellow and purple. Body red, with white ſpots on the under ſide.

SPHINX PORCELLUS: alis integris flavo purpureoque variis, abdomine ſubtus ſanguineo albo punctato. *Linn. Syſt. Nat.* 2. 801. 18.—*Fn. Sv.* 1090.
*Fab. Ent. Syſt. T. 3. p. 1. p.* 373. 52.
*Reſ. Inſ.* 1. *phal.* 1. *tab.* 5.
*Albin. Inſ. tab.* 9.
*Eſp. Inſ.* 2. *tab.* 19.
*Geoff. Inſ.* 2. 88. 12.

A ſpecimen of this Inſect in the winged ſtate was found in Hyde Park this ſummer; it is one of the ſcarceſt of the Britiſh Sphinges, and was found by Harris many years ſince in " meadows—Oſterly " Wood, near Brentford, May 27th."

F 4

The

The larva is of an uniform dull brown, with three eye-fhaped fpots on each fide, and is furnifhed with a tail; it feeds on the epilobium, and changes to Chryfalis about the end of July.

PLATE

# PLATE CCCXV.

## RAPHIDIA OPHIOPSIS.

### NEUROPTERA.

*GENERIC CHARACTER.*

Head depreffed or flat. Mouth armed with two teeth, and furnifhed with four palpi. Three ftemmata. Wings deflected. Antennæ long as the thorax, anterior part of which is lengthened out
and cylindrical. Tail of the female terminated by a flexible crooked
briftle.

*SPECIFIC CHARACTER.*

Thorax cylindrical; a brown marginal fpot on each wing.

RAPHIDIA OPHIOPSIS: thorace cylindrico, alis macula marginali
fufca.
RAPHIDIA OPHIOPSIS. *Linn. Syft. Nat.*
Raphidia notata. *Fab. Spec. Inf.* 1. *p.* 402. 106. *fp.* 1.
       *Roef. Inf.* 3. *tab.* 21. *fig.* 67.
       *Scopoli, Carn.* 711.
       *Schæff. Icon. tab.* 95. 1. 2.
       ———— *Elem. tab.* 107.

We can fcarcely conceive what motive induced Fabricius to confound the Raphidia Ophiopfis and notata as one fpecies in his laft
work, after having defcribed both with accuracy in his former publication: it appears indeed, that his names were erroneous, and his
fynonyms mifapplied; but it was needlefs to correct one error by
committing another.

The

The firft fpecies of *Raphidiæ* known, was figured in the works of Roefel, *Die Fleine Landhaelfige. Lanlibelle fig. 6. 7. pl. 21,* it was defcribed in the Fauna Suecica, by Linnæus, under the fpecific name Ophiopfis; and again in the *Syftema Natura* of the fame author, with a reference to the only figure of it then extant, that of Roefel. Hence it appears that the true Raphidia Ophiopfis of Linnæus, is that figured by this author. In later editions, the works of Sulzer, Schæffer, Geoffroy and Scopoli, were added to the fynonyms, but the figures thus quoted, evidently include two fpecies, one with wings perfectly clear, the other having a marginal black fpot on each. Linnæus feems to have confidered the two as varieties of the fame fpecies, but he is evidently miftaken, for the two fexes of both kinds are now clearly afcertained.

Fabricius has followed Linnæus in his Species Infectorum, has indifcriminately adopted all the fynonyms, and thereby confounded all the figures of the two Raphidæ that have been noticed by authors on European Infects, under the name of Ophiopfis: and after this he defcribes that very fpecies which has fpots on the wings as a new and unfigured kind, under the name of *notata*. Thorace cylindrico alis macula marginali fufca. Habitat in anglia. Had he referred to the volumes of Roefel, he muft have known that his *notata* was the Linnæan Ophiopfis, and if either Infect was new, it muft certainly be that deftitute of fpots.

Gmelin in his Syftema Natura perpetuates the fame error; he follows the Species Infectorum of Fabricius, and gives the characters thus: " R. Ophiopfis alis immaculatis. *Fab.*" & " R. notata, " alis macula marginali fufca. *Fab.*" but whilft the works of Gmelin are preparing for publication, Fabricius alters his opinion; and in the laft work. *Syft. Ent. emendet et aucta,* abolifhes his fpecific cha-racters, and merely fays there is no difference between his former fpecies " Raphidia notata, nullo modo diftincta." *Fab. Ent. Syft. T. 2. p. 99.*

As

# PLATE CCCXV.

59

As we have all the species described by those authors before us, we shall endeavour to restore them to order, and that by retaining the former descriptions of Fabricius, changing the names, and dividing the synonyms, for both are sufficiently characteristic; that with marginal spots is figured by Roesel, Schæffer and Scopoli, and the immaculated or clear-winged kind by Geoffroy and Sulzer. The first we deem the true R. Ophiopsis, and the latter as a distinct insect, which may be called the Raphidia Immaculata with propriety.

Both species of this singular creature are extremely rare. Geoffroy, speaking of the unspotted kind, says he never found it but twice, and then in woods *. The larva is unknown; in the pupa state it is furnished with legs, and runs fast.

* Geoffroy Histoire des Insectes,

PLATE

316

# PLATE CCCXVI.

## PHALÆNA GONONSTIGMA.

### Scarce Vapourer Moth.

### GENERIC CHARACTER.

Antennæ taper from the bafe. Wings in general deflexed when at reft. Fly by night.

### SPECIFIC CHARACTER

#### AND

### SYNONYMS.

Wings incumbent, brown. Two white fpots on the firft wings; one placed on the anterior, and the other nearly oppofite, on the pofterior margin. Female without wings.

PHALÆNA GONONSTIGMA : acis incumbentibus fufcis : maculis duabus albis oppofitis, fœmina aptera. *Linn. Syft. Nat.* 2. 826. 57.—*Fab. Ent. Syft. T.* 3. *p.* 1. *p.* 477. *fp.* 217.
*Roef. Inf.* 1. *phal.* 2. *tab.* 40.
*Albin. Inf. tab.* 90.

---

The Phalæna Gononftigma, and Phalæna Antiqua are very fimilar both in the larva and winged ftate, as well as in the extraordinary appearance of the apterous female. Hence former collectors of Englifh infects denominated them trivially the Scarce and Common Vapourer Moths. It is evident from thofe allufive names, that the

latter

latter was more frequently taken than the other ; at this time Pha‑
læna Antiqua is found very common, but the latter fo rarely, that
we never met with it, in the winged ftate, till this fummer.

Once found the larva on an oak in Coombe Wood, Surry, but it
died foon after.

PLATE

# PLATE CCCXVII.

## PHALÆNA PERSICARIÆ.

### *GENERIC CHARACTER.*

Antennæ taper from the bafe. Wings in general deflexed when at reft. Fly by night.

### *SPECIFIC CHARACTER*

AND

### *SYNONYMS.*

Thorax crefted, wings deflexed, dark and clouded. A white kidney-fhaped fpot, with a yellow lunar pupil in the middle on each.

PHALÆNA PERSICARIÆ: crifta alis deflexis fufco nebulofis: ftig-
mate reniformi albo ; pupilla lunari flava. *Linn.*
*Syft. Nat.* 2. 847. 142.—*Fn. Suec.* 1208.
*Geoff. Inf.* 2. 157. 94.
*Ammir. Inf. tab.* 157.
*Roef. Inf. I. phal.* 2. *tab.* 30.

———————————

A very common infect ; and is often obferved near fruit trees.

PLATE

1

# PLATE CCCXVIII.

## PHALÆNA DIDACTYLUS.

### BIFID-WING PLUME MOTH.

### GENERIC CHARACTER.

Antennæ taper from the bafe.  Wings in general deflexed when at reft.  Fly by night.

* Alucita.  *Linn.*

### SPECIFIC CHARACTER

#### AND

### SYNONYMS.

Wings divided into plumes, brown, barred with white.  Anterior wings confifts of two feathers, pofterior pair of three.

PHALÆNA DIDACTYLUS: alis fiffis fufcus: ftrigis albis anticis
 bifidis, pofticis tripartitis. *Linn. Syft. Nat.* 2. 899.
 454.—*Fn. Sv.* 1453.
Pterophorus Didactylus.  *Fab. Ent. Syft. T.* 3. *p.* 2. *p.* 345. *fp.* 200.
 *Geoff. Inf.* 2. 92. 2.
 *Wien. Verz.* 145. 2.
 *Schæff. Icon. tab.* 93. *fig.* 7.
 *Elem. tab.* 104.

G                                        The

The larva of this very fingular creature is faid to feed on the convolvulus and Geo rivali.  We have confidered it as one of the fcarceft fpecies of the Plume-Moths found in this country : our fpecimen was taken in Epping Foreft, in June  It is a moft beautiful objeƈt for the microfcope.

P L A T E

# PLATE CCCXIX.

## SPHINX LOTI.

### FIVE SPOT BURNET SPHINX.

## *GENERIC CHARACTER.*

Antennæ thickeſt in the middle. Wings deflected when at reſt. Fly ſlow, morning and evening.

## *SPECIFIC CHARACTER*

### AND

## *SYNONYMS.*

Anterior wings greeniſh, with five red ſpots. Poſterior wings red, bordered with fine blue.

ZYGÆNA LOTI : alis anticis viridibus: punctis quinque rubris, poſticis ſanguineis: limbo cyaneo. *Fab. Ent. Syſt.* *T. 3. p. 2. p. 387. ſp. 5.*
SPHINX LOTI. *Wien. Verz.* 45. 3.
  *Schæff. Icon. tab.* 16. *fig.* 6. 7.
Sphinx Loniceræ. *Eſp. Inſ.* 2. *tab.* 24. *fig.* 1.

———————

This beautiful little ſpecies may be eaſily confounded with the Sphinx Filipendula, figured in the ſixth plate of this work; its general reſemblance is ſtriking, and it differs chiefly in the number of red ſpots that adorn the ſuperior wings. Sphinx Filipendula has invariably ſix ſpots on each wing, and the latter as conſtantly only five.

Some

Some readers may be inclined to deem it a mere variety of the fort, from its general appearance, but it will be perceived by the fynonyms quoted above, that all the continental writers on the fub-ject admit it as a diftinct fpecies ; nor can we for a moment hefitate to agree in the fame opinion.

It is rare in this country, and the larva unknown, or at leaft is undefcribed.

PLATE

320

# PLATE CCCXX.

### PAPILIO JANIRA.

MEADOW BROWN BUTTERFLY.

### GENERIC CHARACTER.

Antennæ clubbed at the end. Wings erect when at reft. Fly by day.

### SPECIFIC CHARACTER

AND

### SYNONYMS.

Wings dentated above, brown beneath, firft pair yellowifh, with a black eye-fhaped mark, fecond pair brownifh, with two fmaller eye-fpots.

PAPILIO JANIRA: alis dentatis fufcis; anticis fubtus luteis; ocello utrinque unico, pofticis fubtus punctis tribus. *Linn. Syft. Nat.* 2. 744. 156.—*Fn. Sv.* 1053.—*Fab. Ent. Syft.* 3. *p.* 1. 241. 752. *Schæff. Icon. tab.* 273. *fig.* 1. 2. 5. 6.—*Geoff. Inf.* 2. 49. 17.

β PAPILIO JURTINA: alis dentatis fufcis: anticis fupra litura flava ocello utrinque unico. *Linn. Syft. Nat.* 2. 774. 155.—*Fn. Sv.* 152. *Roeff. Inf.* 3. *tab.* 34. *fig.* 7. 8.

---

Linnæus defcribed the two fexes of this Butterfly as diftinct Species under the names of Janira and Jurtina. The firft is the male and the latter the female infect.

H

The

The larva is hairy, green, with a lateral white line and bifid tail, and feeds on grafs. It is very common in the winged ftate, frequenting meadows, &c. whence it is called the Meadow Brown Butterfly.

PLATE

# PLATE CCCXXI.

## FIG. I. I.

## CHRYSOMELA CORYLI.

### COLEOPTERA.

### *GENERIC CHARACTER.*

Antennæ compofed of globular articulations which become larger towards the ends.

### *SPECIFIC CHARACTER*

### AND

### *SYNONYMS.*

Black. Thorax and wing-cafes teftaceous brown, without fpots.

CHRYSOMELA CORYLI: *Linn. Syft. Nat.* 2. *p.* 598. 88.—*Fn. Suec.* 555.

Cryptocephalus. *Gmel. Linn. Syft. Nat. T.* 1. *p.* 6. 1704. *fp.* 28.

CRYPTOCEPHALUS CORYLI: niger, thorace elytriifque teftaceis immaculatis. *Fab. Spec. Inf.* 1. *p.* 142. *n.* 24.

---

We have frequently obferved this fpecies amongft the Infects of Germany, where it is probably not uncommon. In England it is very rare, having been found only by the Rev. John Burrel of Letherinfet, near Holt, Norfolk. In one fex the thorax is red, in the other black.

H 2

FIG.

# PLATE CCCXXI.

## FIG. II. II.

## CHRYSOMELA SERICEA.

*SPECIFIC CHARACTER*

AND

*SYNONYMS.*

Bluifh green.   Antennæ black.

CHRYSOMELA SERICEA.   *Linn. Syft. Nat.* 2. *p.* 598. *n.* 86.——
*Fn. Sv.* 554.
Cryptocephalus.   *Gmel. Linn. Syft.* 1. *p.* 6. *p.* 1706. *Sp.* 43.


Cryptocephalus fericeus: viridi-cæruleus antennis nigris.   *Fab.*
*Sp. Inf.* 1. *p.* 143. *n.* 32.

Found in June.

322

# 73

# PLATE CCCXXII.

## FIG. I. I.

### PAPILIO ALSUS.

LEPIDOPTERA.

### GENERIC CHARACTER.

Antennæ clubbed at the end, Wings in general erect when at reft. Fly by day.

*Plebeii rurales.*

### SPECIFIC CHARACTER.

Wings entire brown, without fpots; beneath grey, with a row of eye-fhaped fpots.

HESPERIA ALSUS: alis integerrimis fufcis immaculatis fubtus cinereis, ftriga punctorum ocellatorum. *Fab. Ent. Syft. T. 3. p. 1. p. 295. 125. Schæff. Icon. 2. tab. 165. fig. 1. 2.*

———————

This pretty Infect is found late in June. Its larva is unknown.

H 3           FIG.

## FIG. II. II.

## PAPILIO IDAS.

### SPECIFIC CHARACTER.

Wings entire brown. An equal marginal row of red fpots both on the upper and underfide. A black fpot in the middle of the anterior wings.

Papilio Idas: alis integris fufcis, fafcia marginali utrinque rubro-
maculata, anticis macula media nigra.

---

This infect muft not be confounded with the Papilio Idas of Linnæus. The Linnæan P. Idas is evidently the female of P. Argus, a circumftance unknown to that author, who confiders them as a diftinct fpecies from their very diffimilar appearance. In one fex the upper furface is brown, and in the other a fine blue; this is not, however, peculiar to the P. Argus, for feveral of the Papilio tribe known amongft Englifh collectors by the trivial name *Blues* differ in the fame manner.

We fufpect that our Infect has not been defcribed by any author; it is certainly unnoticed by Fabricius in his laft Syftem of Entomo-logy, and the fpecific name *Idas* omitted. This name is therefore preferred for our Infect, which feems to approach nearer to the female Argus defcribed by Linnæus as Idas, than to any other. Found in May.

PLATE

# PLATE CCCXXIII.

## SCARABÆUS LURIDUS.

### GENERIC CHARACTER.

Antennæ terminate in a club, which is divided longitudinally into laminæ or plates.

### SPECIFIC CHARACTER

AND

### SYNONYMS.

Scutellum, thorax and head black. Wing-cafes pale brown, ftriated, and fomewhat teffellated with linear black marks.

Scarabæus luridus: fcutellatus capite tuberculato ater, elytris grifeis nigro ftriatis. *Fab. Ent. Syft. T.* 1. *p.* 29. *Sp.* 91.
Scarabæus luridus. *Oliv. Inf.* 1. 3. 90. 100. *tab.* 18. *fig.* 68. *and tab.* 26. *fig.* 168.
Scarabæus teffellatus. *Myll. Zool. Dan.*
    *Jabl. Coleopt.* 2. *tab.* 18. *fig.* 3.

——————————

Fabricius defcribes this fpecies from a fpecimen in the cabinet of Sir J. Banks, and notes its *habitat* England. From this circum-ftance we may infer that it is not common in other countries, though we are certain it is a native of Denmark and Germany.

H 4                    PLATE

# 324

# PLATE CCCXXIV.

## PHALÆNA SPONSA.

### CRIMSON UNDERWING MOTH.

#### GENERIC CHARACTER.

Antennæ fetaceous. Wing deflexed when at reft. Fly by night.

#### SPECIFIC CHARACTER

##### AND

#### SYNONYMS.

Thorax crefted. Anterior wings greyifh, undulated, fpotted with brown. Pofterior pair crimfon, with two black bars acrofs. Abdomen grey.

PHALÆNA SPONSA : crifta, alis planis cinerafcentibus fufco undulatis: pofticis rubris; fafciis duabus nigris, abdomine undique cinereo. *Linn. Syft. Nat.* 2. 841.
118. *Roef. Inf.* 4. *tab.* 19.

---

In the defcription of Phalæna Nupta, we have offered fome remarks on the Sponfa, Nupta, and Pacta of Linnæus and Fabricius;
and have only to add in this place, that an accurate figure of
P. Pacta is given in *Fuefl. Archiv. tab.* 15. *fig.* 3. This figure is
fmaller than the fpecies found in Great Britain, and in particular
has the upper furface of the abdomen crimfon, as authors have defcribed it.

The

The Synonyms of the three fpecies, as they ftand in the works of Linnæus and Fabricius, are very incorrect. We venture to retain that to Roefel's plate, vol. 4. t. 19. in which the larva we have figured is given.

The Caterpillars feed on the tops of the higheft Oaks, change to the pupa ftate in June, and appears a winged Infect early in the month following.

PLATE

# LINNÆAN INDEX

### TO

## VOL. IX.

---

### COLEOPTERA.

---

### HEMIPTERA.

---

### LEPIDOPTERA.

Papilio

# I N D E X.

## NEUROPTERA.

## DIPTERA.

ALPHA-

# ALPHABETICAL INDEX

TO

## VOL. IX.

# I N D E X.

*This Day is Publiſhed,*
AS A COMPANION
TO THE

## HISTORY of BRITISH INSECTS,

Price Two Shillings and Sixpence each NUMBER, of an entirely new, and
elegantly finiſhed Work,

*To be continued Monthly,*

---

THE
# NATURAL HISTORY
OF
# BRITISH BIRDS;
OR, A
SELECTION OF THE MOST RARE, BEAUTIFUL, AND INTERESTING
B I R D S
WHICH INHABIT THIS COUNTRY:
THE DESCRIPTIONS FROM THE
*SYSTEMA NATURÆ OF LINNÆUS:*
WITH
GENERAL OBSERVATIONS,
EITHER ORIGINAL, OR COLLECTED FROM THE LATEST
AND MOST ESTEEMED
*ENGLISH ORNITHOLOGISTS;*
AND EMBELLISHED WITH
F I G U R E S,
DRAWN, ENGRAVED, AND COLOURED FROM THE ORIGINAL SPECIMENS.

---

BY E. DONOVAN.

---

*LONDON:*
PRINTED FOR THE AUTHOR; AND FOR F. AND C. RIVINGTON,
No. 62, ST. PAUL's CHURCH-YARD. 1798.

---

## C O N D I T I O N S.

I. THIS Work will be compriſed in Sixty-two Numbers, *Price Two
Shillings and Six-pence each Number.*

II. Two Plates will be given in each Number. They will be taken on
WHATMAN's Superfine Wove, or Vellum Drawing Paper, *and
finiſhed in a peculiar Style of Elegance, from Original Specimens now
in the Collection of the* AUTHOR.
The Letter Preſs will be on the fineſt Wove Printing Paper, and
Hot-preſſed.

III. An Index will be given in every twelfth Number; and the Work
will form Five handſome Volumes in Royal Octavo.

# ADDRESS.

THE liberal Patronage with which the HISTORY of BRITISH INSECTS has been honoured, has induced the Proprietors to extend their Views, and refpectfully to folicit the Encouragement of the Public to a further Difplay of the Natural Productions of our Native Country. They have determined, under the Title of a NATURAL HISTORY OF BRITISH BIRDS, to produce an elegantly finifhed Collection of Plates of the moft interefting among thofe which inhabit this Country. This Work will be an handfome Companion to their ENTOMOLOGY; and, to render it an acceptable Acquifition, as well to the Man of Science as the Amateur of Natural Hiftory, the Linnæan Defcriptions will be united with other Information.

This Defign cannot, it is hoped, fail to meet with public Approbation and Encouragement; for though there are already feveral valuable Works including this Divifion of Zoology, yet they are of fuch Expence as to exclude all Purchafers except the very Affluent; while this will offer to a much more general Clafs of Readers an elegant NATURAL HISTORY OF BRITISH BIRDS.

The Proprietors being refolved to execute the Work with Accuracy and Elegance, have, at a very confiderable Expence, collected the living, or preferved Specimens of all the BIRDS intended for the Publication, whence the Figures will be drawn, engraved, and coloured.

In the firft Contemplation of this Work, the Proprietors intended to have produced a complete Illuftration of all the Birds that inhabit this Ifland, amounting together to more than 250 Subjects: But confidering the Extent of fuch a Production, they have fince preferred giving Figures only of thofe that are moft remarkable, beautiful, or rare: Their Subfcribers having, however, after its Completion, been difpofed to offer further Encouragement, a concife SUPPLEMENT will be added.

The Proprietors, though well affured that they might fay much more in Praife of the propofed Execution of this Work, without exceeding the Truth, will not hazard the Appearance of Exaggeration. They prefer the Approbation which will undoubtedly attend the actual Merit of Performance, to any Eagernefs of Expectation in the Public which their Promifes might raife.

N. B. The whole Work, being printed off, may be had complete, in five Volumes. Price in Boards, 7l. 15s.

# THE

# NATURAL HISTORY

OF

# BRITISH INSECTS;

EXPLAINING THEM
IN THEIR SEVERAL STATES,

WITH THE PERIODS OF THEIR TRANSFORMATIONS,
THEIR FOOD, OECONOMY, &c.

TOGETHER WITH THE

## HISTORY OF SUCH MINUTE INSECTS

AS REQUIRE INVESTIGATION BY THE MICROSCOPE.

THE WHOLE ILLUSTRATED BY

# COLOURED FIGURES,

DESIGNED AND EXECUTED FROM LIVING SPECIMENS.

———

By E. DON'OVAN.

———

VOL. X.

———

LONDON:

PRINTED BY BYE AND LAW, ST. JOHN'S SQUARE, CLERKENWELL,

FOR THE AUTHOR,

And for F. and C. RIVINGTON, No 62, ST. PAUL'S CHURCH YARD.

MDCCCI.

# THE
# NATURAL HISTORY
OF
# BRITISH INSECTS.

---

## PLATE CCCXXV.

### SPHINX TILIÆ.

#### LIME HAWK MOTH.

#### LEPIDOPTERA.

### *GENERIC CHARACTER.*

Antennæ thickeſt in the middle. Wings deflexed when at reſt. Fly ſlow morning and evening.

### *SPECIFIC CHARACTER*

#### AND

### *SYNONYMS.*

Wings angulated, greeniſh clouded with brown, two triangular olive ſpots diſpoſed as a bar acroſs the anterior wing, tips white. Poſterior wings yellow brown with a tranſverſe dark bar.

SPHINX TILIÆ : alis angulatis vireſcenti nebuloſis ſaturatius faſciatis,
poſticis ſupra luteo teſtaceis. *Linn. Syſt. Nat.* 2.

A 2

797.

# PLATE CCCXXV.

**4**

797. 3.—*Fn. Sv.* 1085.—*Fab. Ent. Syft. T.* 3. *p.* 1. *p.* 358. *Sp.* 10.

*Albin. Inf. tab.* 10.

*Roef. Inf.* 1. *phal.* 1. *tab.* 2.

*Schæff. Elem. tab.* 116. *fig.* 1.

*Schæff. Icon. tab.* 100. *fig.* 1. 2.

*Merian. Europ.* 2. *tab.* 24.

*Efp. Inf.* 2. *tab.* 3.

*Geoffr. Inf.* 2. 80. 2.

———————

The larva of this elegant Infect feeds on the Lime tree. In September it changes to the pupa, and the Sphinx is produced in May.—It is very common in moft parts of the country.

PLATE

326

# 5

# PLATE CCCXXVI.

## PTINUS PECTINICORNIS.

### COLEOPTERA.

### *GENERIC CHARACTER.*

Antennæ filiform, the laft articulation longeft. Thorax roundifh, with a margin into which the head is drawn back.

### *SPECIFIC CHARACTER.*

Brown. Antennæ yellowifh and pectinated.

PTINUS PECTINICORNIS: fufcus antennis luteis pectinatis. *Linn. Syft. Nat. p. 1.*

---

This fpecies differs very much from the other Infects of the Ptinus genus in having feathered antennæ. Geoffroy, who defcribed it before Linnæus, called it Ptilinus. Linnæus placed it in his Syftem in the Ptinus genus, and to diftinguifh it named it fpecifically pectinicornis. The two fexes of this Infect may be diftinguifhed by the form of the antennæ ; thofe of the female are but flightly pectinated, that which we have figured is the male having large feather d antennæ.

This creature lives in decayed wood.

A 3                                    PLATE

# PLATE CCCXXVII.

## FIG. I.

### PHALÆNA LEPORINA.

MILLER OF MANSFIELD MOTH.

LEPIDOPTERA.

*GENERIC CHARACTER.*

Antennæ taper from the bafe. Wings in general deflexed when at reft. Fly by night.

*SPECIFIC CHARACTER*

AND

*SYNONYMS.*

Wings deflexed, white, fprinkled with forked black fpots: no fpots on the abdomen.

PHALÆNA LEPORINA : alis deflexis albis : punctis ramofis, abdomine inmaculato. *Linn. Syft. Nat.* 2. 838. 9. *Fn. Sv.* 1176.—*Fab. Ent. Syft. T.* 3. *p.* 1. *p.* 453. 144. *Degeer Inf.* 1. *tab.* 12. *fig.* 10. 11. 17. *Fyefl. Magaz.* 2. *tab.* 1. *fig.* 1—3.

---

We cannot account for the very abfurd name Englifh Aurelians have given to this Infect. It probably originated in fome trivial event, which has been long fince forgotten, but as the Infect will be

A 4        better

better known by that name than any other we could adopt, it is thought moft advifable to retain it.

This Moth is uncommonly fcarce in Great Britain. The larva is of a pale or greenifh white colour with three longitudinal ftripes of brown and a few black bars acrofs. It feeds on the Willow and Alder. Four of thofe larvæ were taken in the wood of Darent, 1793 ; and one of them was reared to the winged ftate.

## F I G. II.

## P H A L Æ N A   A L N I.

*SPECIFIC CHARACTER*

AND

*SYNONYMS.*

Thorax crefted. Anterior wings brown with two broad fpaces of grey, divided by a tranfverfe dark bar, and a kidney-fhaped fpot in the middle. Pofterior wings whitifh with a marginal row of brown fpots, pale brown at the apex.

PHALÆNA ALNI: criftata alis deflexis fuliginofis : areis duabus cineraſcentibus priore puncto marginali nigro. *Linn. Syft. Nat.* 2. 845. 134.—*Fab. Ent. Syft.* *T.* 3. *p.* 2. *fp.* 89.
Noctua Degener.　*Wien. Verz.* 70. 4.—*Degeer. Inf.* 1. *tab.* 11. *fig.* 25. 28.

One of the rare Englifh fpecies of Phalænæ known amongft collectors by the general name of Portland Moths, having been firft difcovered and introduced to notice as natives of this country by the late Dutchefs Dowager of Portland.

F I G.

## F I G  III.

### PHALÆNA DIPSACEA.

#### *SPECIFIC CHARACTER*

AND

#### *SYNONYMS.*

Thorax fmooth. Anterior wings pale clay colour, with a broad brown ramofe bar acrofs. Pofterior pair black, with an irregular pale oblique bar, and double fpot of the fame near the pofterior margin.

PHALÆNA DIPSACEA: lævis alis deflexis pallidis: fafcia lata fufca, pofticis albo nigroque variis. *Lin. Syft. Nat.* 2. 856. 185.—*Fab. Ent. Syft. T.* 3. *p.* 2. *p.* 33. *fp.* 83.
*Wien. verz.* 89. 3.

Hitherto confidered as a fcarce or at leaft very local fpecies. Dr. Latham found it in great abundance in a clover field near Dartford, Kent. The larva is defcribed; it is red with broken or interrupted white lines and a cinereous head. It feeds on the Centaurea, Plantain and Tragopogon.

P L A T E

# 328

# PLATE CCCXXVIII.

## PHALÆNA FAGI.

### Lobster Moth.

### Lepidoptera.

## GENERIC CHARACTER.

Antennæ taper from the bafe. Wings in general deflexed when at reft. Fly by night.

## SPECIFIC CHARACTER

### AND

## SYNONYMS.

Wings reverfed, reddifh afh colour, with two incurvated yellowifh lines acrofs the firft pair.

PHALÆNA FAGI: alis reverfis rufo cinereis: fafciis duabus linearibus luteis flexuofis. *Linn. Syft. Nat.* 2. 816. 30.
— *Fn. Sv.* 113. — *Fab. Ent. Syft. T.* 3. *p.* 1. *p.* 422. *fp.* 51.
*Albin. Inf. tab.* 58.
*Wien. Verz.* 63. 2.
*Roef. Inf.* 3. *tab.* 12.
*Act. holm.* 1749. 132. *tab.* 4. *fig.* 10. 14.

The trivial name of Lobfter Moth, which this fpecies has acquired from the fingular form of its larva, cannot be unfamiliar to the Englifh Aurelian, though the Moth itfelf is in the poffeffion of few. The larva

was

was figured and defcribed by Albin, and collectors about the middle
of the laft century occafionally met with it in the woods near London, which have been fince deftroyed. At that time it was however
fcarce, and being difficult to rear, the Moth has always been deemed
one of the moft valuable Britifh fpecies of the Lepidoptera tribe.

An old collector at Hoxton once informed us, that the larva of this
Infect was called the BREECHES Caterpillar about fifty years ago ;
that it was in great requeft by moft collectors of his time, and that
he deemed himfelf fortunate in finding two fpecimens of it in the
courfe of his life, though he had not reared either. Thofe were
taken on fome Cheftnut trees which grew at that time in St. George's
fields. The late Mr. Bentley found it once on the Beech, and Mr.
Francillon has a fpecimen of it in his cabinet, which he met with
himfelf. Our figures are copied from Mr. Francillon's fpecimen,
and the drawings Roefel has given of it in his Hiftory of the Infects
of Germany.

PLATE

# 329

# PLATE CCCXXIX.

## SPHEX APPENDIGASTER.

### SMALL-BODIED ICHNEUMON WASP.

### HYMENOPTERA.

### GENERIC CHARACTER.

Mouth armed with jaws, no tongue. Antennæ confift of ten articulations. Wings extended, without folds, and laid horizontally upon the back. Sting fharp and pointed, and concealed within the abdomen.

### SPECIFIC CHARACTER

#### AND

### SYNONYMS.

Black. Abdomen fmall, joined to the thorax by a footftalk. Pofterior legs very long.

SPHEX APPENDIGASTER: atra abdomine petiolata breviffimo, pedibus pofticis longiffimis. *Linn. Syft. Nat. 2.
945. 12.—Gmel. Linn. Syft. Nat. p. 2723. 245.
fp. 12.*

EVANIA APPENDIGASTER: atra abdomine petiolato breviffimo
dorfo thoracis impofito, pedibus pofticis longiffimis. *Fab. Ent. Syft. T. 2. 141. 1.
Degeer. Inf. 3. 394. tab. 30 fig. 14.
Reaum. Inf. 6. tab. 31. fig. 13.*

The

The novelty of this creature will be immediately obvious to thofe in the flighteft degree acquainted with the hymenopterous tribes of Infects. At firft fight it has the exact appearance of an Infect deprived of the body, for the abdomen is extremely fmall in proportion to the other parts, and fo much recurved or bent under the pofterior part of the thorax as to be fcarcely vifible.

It is not more remarkable for its fingularity than rarity, for we have ventured to introduce it as an Englifh Infect on one authority only.—A fpecimen of it was lately taken by the Rev. James Coyte of Ipfwich, in Suffolk. We have it from the South of Europe.

PLATE

# 330

# PLATE CCCXXX.

## PHALÆNA ACERIS.

### SYCAMORE MOTH.

### LEPIDOPTERA.

## *GENERIC CHARACTER.*

Antennæ taper from the base. Wings in general deflexed when at rest. Fly by night.

## *SPECIFIC CHARACTER*

### AND

### *SYNONYMS.*

Thorax crested. Wings deflexed, grey, undulated with black, and a black dagger-like mark at the base of the anterior pair.

PHALÆNA ACERIS: cristata alis deflexis canis nigro undatis, abdomine subtus basi brunneo. *Linn. Syst. Nat.* 2. 846. 137.—*Fn. Sv.* 1179.—*Fab. Ent. Syst. T.* 3. *p.* 2. *p.* 107. *sp.* 322.
*Wilks pap.* 32. *tab.* 2. *c.* 6.
*Reaum. Inf.* 1. *tab.* 34. *fig.* 11.
*Frisch. Inf.* 1. *tab.* 5.

The larva of this species feeds on the Sycamore, it becomes a pupa late in August, and appears in the winged state in June.

Another

Another Moth very analogous to this species is known amongst English collectors by the name of Sycamore likeness; it is exceedingly similar in its colour and marks, but is destitute of the small dagger-form black spot which is situated near the base of the anterior wings in Phalæna Aceris.

PLATE

# 331

# PLATE CCCXXXI.

## PHALÆNA DELPHINII.

### Pease-blossom Moth.

### Lepidoptera.

#### GENERIC CHARACTER.

Antennæ taper from the bafe. Wings in general deflexed when at reft. Fly by night.

#### SPECIFIC CHARACTER

AND

#### SYNONYMS.

Thorax crefted. Wings deflexed, firft pair purple with two broad tranfverfe whitifh bars, fecond pair pale brown.

PHALÆNA DELPHINII: criftata alis deflexis purpurafcentibus: faf_
ciis duabus albidis, pofticis obfcuris. *Linn. Syft.*
*Nat.* 2. 857. 188.—*Fab. Ent. Syft. T.* 3. *p.* 2.
*p.* 90. *fp.* 267.
*Geoff. Inf.* 2. 164. 109.
*Merian. Europ.* 1. *tab.* 40.
*Roef. Inf.* 1. *phal.* 2. *tab.* 12.
*Panz. Fn. Germ.* 7. *tab.* 17.

---

The Phalæna Delphinii is extremely rare. A traditionary opinion feemed to prevail amongft the old collectors of Englifh Infects,

B                                         that

that it had been taken in this country, but the fact was not clearly ascertained till within the last two years.

The late Duchess of Portland, it is reported, once found a mutilated wing of some Phalæna hanging in a cobweb, which it was conjectured had belonged to this species; but on such slender authority few were disposed to confider it as a British Insect: and thence it remained a subject of dispute till the summer of 1799, when our worthy friend W. Jones, Esq. met with a charming specimen of it alive in his own garden at Chelsea; and thereby removed every doubt respecting it, as a British Species.

The larva feeds on the Larkspur, and is figured by Roesel together with the eggs and pupa; those figures we have copied in the annexed plate, as they render the history of this interesting Insect more complete than our limited information would otherwise permit. —The larva seems to bear some resemblance to those of Phalæna Verbasci, or Water Betony Moth; the colours are nearly the same, but the black spots in the former are more numerous.

PLATE

# 332

# PLATE CCCXXXII.

## BLATTA LAPPONICA.

### HEMIPTERA.

### *GENERIC CHARACTER.*

Head inflected. Antennæ fetaceous. Elytra femicoriaceous. Thorax flat, orbicular and margined. Feet formed for running.

### *SPECIFIC CHARACTER*

### AND

### *SYNONYMS.*

Yellowifh. A few black fpots on the longitudinal ridge of the wing cafes.

BLATTA LAPPONICA: flavefcens elytris nigro maculatis. *Linn. Fn. Sv.* 863.—*Gmel. Linn. Syft. Nat. p.* 2044. *Sulz. Inf. t.* 8. *f.* 3. *Geoffr. Inf. par.* 1. *p.* 381. *n.* 3.

---

We believe this is a fcarce Species in Great Britain. The late Mr. Bentley has taken it about Epping.—It is very common in Lapland.

# 333

# PLATE CCCXXXIII.

## FIG. I.

## PHALÆNA REPANDARIA.

### THE MOTTLED BEAUTY.

#### GENERIC CHARACTER.

Antennæ taper from the bafe. Wings in general deflexed when at reft. Fly by night.

#### SPECIFIC CHARACTER.

Antennæ pectinated. Wings grey, undulated, clouded with brown, and furrounded with a black waved marginal line.

PHALÆNA REPANDARIA: pectinicornis, alis cinereis: omnibus fufco-undatis; pofticis margine repando atro.
Phalæna repandata. *Linn. Syft.* 866. 235.—*Fn. Suec.* 1260. *Kleeman Inf.* 1. *t.* 14. *fig.* 1. 2. *t.* 28. *f.* 1.

———————

The two Moths figured I. I. in the annexed plate are fuppofed to be the male and female of the fame fpecies. Taken in June, about Willows.

C

FIG.

## FIG. II.

## PHALÆNA CONSORTARIA.

### THE PALE OAK BEAUTY.

### *SPECIFIC CHARACTER.*

Antennæ feathered. Wings pale greyiſh, waved with brown. An eye-ſhaped ſpot, whith an oblong white pupil in the middle of the poſterior wings.

PHALÆNA CONSORTARIA: pectinicornis alis dentatis griſeis fuſco
         ſtrigoſis: poſticis puncto ocellari oblongo albido.
         *Fab. Ent. Syſt.* 3. *b.* 137. 29.

———————————

Found on the Oak in June.

PLATE

334

# PLATE CCCXXXIV.

## PHALÆNA FLAVOCINCTA.

### GREAT RANUNCULUS MOTH.

### *GENERIC CHARACTER.*

Antennæ taper from the bafe. Wings in general deflexed when at reft. Fly by night.

### *SPECIFIC CHARACTER*

### AND

### *SYNONYMS.*

Thorax crefted. Wings deflexed: margins dentated. Firft pair greyifh brown, obfcurely clouded and variegated with fmall orange-yellow fpots.

PHALÆNA FLAVOCINCTA: criftata alis deflexis dentatis fufco
     cinereoque variis fulvo punctatis. *Fab. Ent. Syft.*
     *T. 3. p. 2. p.* 114. *fp.* 334.
Noctua flavocincta. *Wien. Verz.* 72. 2.
     *Roef. Inf.* 1. *phal.* 2. *tab.* 54. 55.

---

The larva is fuppofed to feed on the black Cherry and Sloe; on the latter of which we once found it. Its Englifh name implies that it feeds alfo on fome plant of the Ranunculus genus. In the winged ftate it is fometimes met with in gardens.

Another Infect analagous to this fpecies has been named the fmall Ranunculus Moth.

C 2

PLATE

# 335

# PLATE CCCXXXV.

## CHRYSOMELA MARGINELLA.

### GENERIC CHARACTER.

Antennæ compofed of globular articulations, increafing in bulk towards the ends.   Thorax and elytra without margins.

### SPECIFIC CHARACTER.

Black.   Head, thorax, feet, and exterior border of the wing-cafes yellowifh.

CHRYSOMELA MARGINELLA ; niger, capite thorace pedibus co-leoptrorumque limbo flavis.

---

A new fpecies, taken in Coombe Wood in the month of July, by Alex. M'Leay, Efq. and the Rev. Mr. Kirby.

C 3                    PLATE

# 336

# PLATE CCCXXXVI.

## PHALÆNA PINIARIA.

### PINE MOTH.

### *GENERIC CHARACTER.*

Antennæ taper from the bafe. Wings in general deflexed when at reft. Fly by night.

### *SPECIFIC CHARACTER*

AND

### *SYNONYMS.*

Antennæ feathered. Upper fide brown, with broad yellowifh fpots in the difk. Under fide mottled, and clouded with two dark bars acrofs the pofterior pair.

PHALÆNA PINIARIA: pectinicornis alìs fufcis flavo maculatis fubtus nebulofis: fafciis duabus fufcis. *Linn. Syft. Nat.* 2. 861. 210.
*Fn. Sv.* 1233.
*Fab..Ent. Syft. T.* 3. *p.* 2. *p.* 141. *fp.* 45.
*Clerk. phal. tab.* 1. *fig.* 10.
*Schæff. Icon. tab.* 159. *fig.* 1. 2.

This rare and curious fpecies of Phalæna has been for fome years admitted to the cabinets of Englifh Natural Hiftory, but on the moft dubious authority. That it is an inhabitant of Great Britain, is

C 4        however

however now afcertained, for in the fummer of the prefent year: about the laft week in June, it was obferved in great plenty in a fir wood at Crathis, on the north bank of the river Dee, in Mearn- fhire, Scotland, by George Milne, Efq. of Surrey Place, Walworth. They feldom fly low, and are confequently taken with fome difficulty.

The larva is green, ftreaked with white and yellow, and feeds on the Pine. Linnæus and Fabricius add the Lime and Alder alfo.

PLATE

# 337

# PLATE CCCXXXVII.

## LIBELLULA VULGATA.

### Common Dragon Fly.

### *GENERIC CHARACTER.*

Mouth armed with more than two jaws. Antennæ fhorter than the thorax. Wings expanded without folds. Tail of the male furnifhed with forceps.

### *SPECIFIC CHARACTER*

### AND

### *SYNONYMS.*

Wings tranfparent. Abdomen cylindrical and reddifh.

Libellula Vulgata: alis hyalinis, corpore cylindrico rufo. *Linn. Syft. Nat.* 2. 901. 3.—*Fn. Sv.* 1461. *Roef. Inf.* 2. *Aquatic.* 2. *tab.* 8.

———————

This is the moft abundant fpecies of the Libellula tribe; frequenting ditches and other watery places during fummer. The colours in different fpecimens vary exceedingly.

FIG.

## FIG. II.

### LIBELLULA GRANDIS.

#### SPECIFIC CHARACTER.

Wings yellowish. Abdomen cylindrical, variegated, four yellow lines on the thorax.

Libellula fulva alis flavefcentibus, thoracis lateribus lineis duabus flavis, fronte flavefcente cauda diphylla. *Geoff. Inf.* 2. 227. 12.

LIBELLULA GRANDIS: alis glaucefcentibus thoracis lateribus lineis quatuor flavis. *Linn. Syft. Nat.* 2. 903. 9.—*Fn. Sv.* 1467.

Libellula fufca capite rotundato, thorace lineolis quatuor tranfverfis luteis, alis flavicantibus, abdomine cylindrico. *Degeer. Inf.* 2. 2. 45. *tab.* 20. *fig.* 6.

―――――――

When the fine fpecies of Libellula *grandis* was figured in plate 166 of this work, we were not in poffeffion of the variety with yellow wings which Linnæus defcribes. It has fince been difcovered in the neighbourhood of Batterfea, and we deem it too interefting to be omitted.

The yellow colour of the wings can by no means induce us to think it a diftinct fpecies from the variety with wings perfectly tranfparent, efpecially as the fame variation is obferved in moft other fpecies of the fame tribe, and particularly amongft the Englifh fpecies in flaveola virgo and puella. The marks and colours of the abdomen and thorax are ftill more liable to variation, and can fcarcely furnifh any precife character for a fpecific difference.

PLATE

# 338

# PLATE CCCXXXVIII.

## FIG. I.

## PHALÆNA CUCULLA.

MAPLE PROMINENT MOTH.

### *GENERIC CHARACTER.*

Antennæ taper from the bafe. Wings in general deflexed when at reft. Fly by Night.

### *SPECIFIC CHARACTER.*

Tongue fpiral. Thorax crefted. Wings deflexed, margin denticulated, yellow brown clouded with ferruginous and marked obliquely with feveral interrupted parallel and interwoven waved ftreaks. A broad white band next the exterior margin.

PHALÆNA CUCULLA : fpirilinguis, criftata alis deflexis denticulatis ochraceis maculis ferrugineis, fafciaque marginali albida ftriis intertexta fufcis. *Linn. Syft. Nat.* 81.

---

This fpecies is uncommonly rare, and has, we believe, not been figured by any author, unlefs *fig.* 1. *tab.* 71. *of Efper* is intended for the fame infect.—It feeds on the maple.

FIG.

## F I G.  II.

### PHALÆNA RUBAGO.

#### *SPECIFIC CHARACTER.*

Anterior wings yellow : bafe, coftal fpot, and oblique broad bar near the apex reddifh, fprinkled with points of a darker colour.

PHALÆNA RUBAGO : alis anticis flavis ; bafi macula coftali, fafcia lata obliqua punctifque ferrugineis.

───────────────

A new and undefcribed Britifh fpecies.  Once found in the wood at Hornfey.

────────────────────────────

## F I G.  III.

### PHALÆNA PAR.

#### KITTEN-LIKENESS MOTH.

#### *SPECIFIC CHARACTER.*

Anterior wings greyifh white, with a broad clouded bar acrofs the middle.  Pofterior wings darkeft near the exterior margin.

PHALÆNA PAR: alis anticis grifeo-albidis: fafcia lata nebulofa, pofticis extus fufcentibus.  *Marfh. Mfs.*

───────────────

Sometimes found fticking againft walls and trunks of trees, and is certainly an hitherto undefcribed fpecies.

P L A T E

# 339

# PLATE CCCXXXIX.

## PHALÆNA SULPHURALIS.

### MARGATE BEAUTY.

#### GENERIC CHARACTER.

Antennæ taper from the bafe. Wings in general deflexed when at reſt. Fly by night.

#### SPECIFIC CHARACTER

##### AND

#### SYNONYMS.

Firſt wings yellow, with irregular connected ſtreaks of black ſpots, and detached marks of the ſame colour on the anterior margin. Second pair brown.

Phalæna Sulphuralis. *Linn. Syſt. Nat.* 2. 881. 333.
BOMBYX LUGUBRIS : alis deflexis flavis: rivulis punctiſque atris, poſticis fuſcis. *Fab. Ent. Syſt. T.* 3. *p.* 1. *p.* 467. *ſp.* 188.—*Schæff. Icon. tab.* 9. *fig.* 14. 15.

———————

A ſcarce inſect in this country. It is ſaid to have been firſt diſcovered at Margate, and from that circumſtance was afterwards known amongſt Engliſh collectors by the name of *Margate beauty.*— Another inſect, by no means ſimilar, has however received the ſame name, having been met with at Margate likewiſe.

FIG.

## F I G. II.

### PHALÆNA PURPURALIS.

PURPLE AND GOLD MOTH.

*SPECIFIC CHARACTER*

AND

*SYNONYMS.*

Purple, with two irregular yellow bands continued acrofs both the upper and under wings.

PHALÆNA PURPURALIS: alis purpurafcentibus: omnibus fafciis duabus luteis. *Linn. Syft. Nat.* 2. 883. 342.— *Fn. Sv.* 1356.

Found on nettles in May.

PLATE

# 340

# PLATE CCCXL.

## FIG. I.

### PHALÆNA LAPPÆ.

#### THE BURDOCK MOTH.

### *GENERIC CHARACTER.*

Antennæ taper from the bafe. Wings in general deflexed when at reft. Fly by night.

### *SPECIFIC CHARACTER.*

Firft wings ferrugineous; a broad yellow clouded bar with three yellow eye-fhaped fpots acrofs the middle. A yellow fpot at the bafe, and another at the apex. Second wings pale, with an obfcure tranfverfe ftreak.

PHALÆNA LAPPÆ: alis ferrugineis: bafi ftigmatibus fafcia macu-laque apicis flavis fufco-nebulofis, pofticis pallidis ftriga obfcura. *Marfh. Mfs.*

It feeds on the Burdock.

FIG.

## FIG. II.

### PHALÆNA CITRINA.

*SPECIFIC CHARACTER.*

Thorax crested. First wings yellowish, with two transverse bands of brown; the interior one interrupted and enclosed between two irregular whitish streaks.

PHALÆNA CITRINA alis flavescentibus, lineis duabus irregularibus
transverfis albidis, fasciis duabus fuscis, interiore
interrupta.

Ernft, in the Papillons de l'Europe, *fig.* 378. gives the figure of a Phalæna not very diffimilar to our species, and probably a mere variety of it. Under this idea we have named it Citrina, from the French name La Citrinne, adopted by Ernft, for it does not appear to have been either figured or described by any other author.

This choice and beautiful Infect is one of those discovered by the late Duchefs of Portland, and is in the Cabinet of Mr. Francillon.

## FIG. III.

### PHALÆNA CLAVIS.

*SPECIFIC CHARACTER.*

First wings reddish, with a paler dash along the middle, ending near a kindney eye-shaped mark and a contiguous white spot. An interrupted dark mark at the base.

PHALÆNA

# PLATE CCCXL.     37

Phalæna Clavis: alis fufco cinereis linea media punctoque albis,
lineola interrupta bafeos maculifque fufcis.

———————

Several varieties of this Infect appear to have been figured in the
work of Ernft already quoted, but none of them agree fo precifely
with our fpecimen as to permit us to refer to his plates.—We believe it is not defcribed by Fabricius or any other fyftematic author.

D                    PLATE

# 341

# PLATE CCCXLI.

## BLATTA GERMANICA.

### GENERIC CHARACTER.

Antennæ fetaceous. Head inflected. Thorax flat, orbicular, and margined. Abdomen terminated in two appendices. Feet made for running.

### SPECIFIC CHARACTER

#### AND

### SYNONYMS.

Livid brown, with two black parallel lines on the thorax.

BLATTA GERMANICA : livida thorace lineis duabus parallelis nigris. *Linn. Syst. Nat.* 2. 688. 9.—*Fab. Ent. Syst. T.* 2. *p.* 10. *sp.* 220. *Herbst. Arch. tab.* 49. *fig.* 10.

--------

Abundant in Germany, but very rare in this country.

PLATE

# 342

# PLATE CCCXLII.

## FIG. I.

## PHALÆNA LICHENARIA.

### *GENERIC CHARACTER.*

Antennæ taper from the bafe. Wings in general deflexed when at reft. Fly by night.

### *SPECIFIC CHARACTER*

#### AND

### *SYNONYMS.*

Antennæ feathered, wings variegated with green and grey, and marked acrofs with two black lines, the interior one recurved, and the exterior waved and bent in the oppofite direction.

PHALÆNA LICHENARIA: pectinicornis alis viridi cinereoque variis: ftrigis duabus nigris; anteriore recurva,
pofteriore undato flexuofa. *Fab. Ent. Syft. T.* 3.
*p.* 2. *p.* 145. *fp.* 59.

The two fexes of this Infect is figured in the annexed plate, *fig.* 1. 1.—The pectinated antennæ denote the male.

FIG.

## F I G. II.

## PHALÆNA SPARTIATA.

### Broom Moth.

*SPECIFIC CHARACTER*

AND

*SYNONYMS.*

Antennæ fetaceous. Firſt wings deep grey with large oblong brown ſpots, encircled with white. Second wings aſh-coloured.

Phalæna Spartiata : feticornis alis oblongis fuſcis : vitta albida, poſticis cinereis. *Fab. Ent. Syſt. T.* 3. *p.* 2. *p.* 181. *ſp.* 188.
Phalæna Spartiata. *Fueſl. Arch.* 2. *tab.* 5.

———————————

Feeds on the broom, and is found in the winged ſtate in June and July.—It is ſcarce, or at leaſt very local.

F I G.

## FIG. III.

## PHALÆNA UNDULATA.

### Scallop-shell Moth.

*SPECIFIC CHARACTER*

AND

*SYNONYMS.*

Antennæ fetaceous. Wings pale, and uniformly ftreaked tranf-verfely with numerous fcalloped lines of brown.

Phalæna Undulata: feticornis alis omnibus ftrigis confertif-
fimis undulatis fufcis. *Linn. Syft. Nat.* 2. 867.
239.—*Fn. Sv.* 272.
*Clerk. Inf. tab.* 6. *fig.* 3.
*Harr. Inf. tab.* 2. *fig.* 5. 6.

———————

Sometimes taken in Kent, particularly in Darent Wood, Dartford. It feeds on the Willow and Oak, and appears in the winged ftate in June.

PLATE

343

# PLATE CCCXLIII.

## FIG. I.

## PHALÆNA JANTHINA.

### ORANGE UNDERWING MOTH.

### GENERIC CHARACTER.

Antennæ taper from the bafe. Wings in general deflexed when at reft. Fly by Night.

### SPECIFIC CHARACTER

### AND

### SYNONYMS.

Thorax crefted. Firft wings incumbent, grey, variegated with brown, and marks of white. Second pair black with a large orange fpot in the middle, and an exterior border of the fame colour.

PHALÆNA JANTHINA: criftata alis incumbentibus grifeis: litura albida, pofticis atris: macula media margineque ferrugineis. *Fab. Ent. Syft. T.* 3. *p.* 2. *p.* 59. *fp.* 166.
Phalæna Janthina. *Wien. Verz.* 78. 9.
Phalæna Domiduca. *Fuefl. Arch.* 3. *tab.* 16.

———————————————

Found in Darent Wood in the month of July. The larva is faid to be white, with undulated brown ftreaks, and fpotted next the pofterior part with black.

F                                                FIG.

# FIG. II.

## PHALÆNA ORBONA.

### PALE YELLOW UNDERWING MOTH.

*SPECIFIC CHARACTER*

AND

*SYNONYMS.*

Thorax crested. First wings incumbent, liver-colour, with obscure spots. Second wings yellow, with a brown lunar mark in the middle, and a band of the same colour near the posterior margin.

PHALÆNA ORBONA : cristata alis incumbentibus hepaticis : posticis flavis ; lunula strigaque postica fuscis. *Fab. Ent. Syst. T. 3. p. 2. p. 57. sp. 158.*

This species is far less frequent than the Phalæna Pronuba, to which, at first glance, it bears some resemblance. Fabricius describes it as a native of Germany, and it does not occur in the Works of any Author as a British Insect.

FIG.

PLATE CCCXLIII.     47

# FIG. III.

## PHALÆNA ARBUTI.

### SPECIFIC CHARACTER.

Thorax crefted. Wings deflexed, brown; pofterior pair black, with a yellow band acrofs the middle.

PHALÆNA ARBUTI: criftata alis deflexis fufcis: pofticis nigris flava. *Fab. Ent. Syft. T.* 3. *p.* 2. *p.* 126. *fp* .380.

———————

Defcribed by Fabricius in his laft Work as an Englifh Infect, from which we may infer, that it is not common in other parts of Europe.

F 2        PLATE

# 344

# PLATE CCCXLIV.

## JULUS COMPLANATUS.

### GENERIC CHARACTER.

Feet numerous. Twice as many on each fide as the fegments of the body. Antennæ moniliform. Palpi two, articulated, body femicylindrical.

### SPECIFIC CHARACTER

#### AND

### SYNONYMS.

Antennæ clavated. Body flat. Tail acute.

JULUS COMPLANATUS : pedibus utrinque 30, corpore planiufculo, antennis clavatis. *Linn. Syft. Nat.* 2. 1065. 4. —*Fn. Sv.* 2068.

JULUS COMPLANATUS : pedibus utrinque 30, corpore planiufculo, cauda acuta. *Fab. Ent. Syft.* 2. *p.* 393.

---

Fabricius as well as Linnæus confiders the number of feet as an effential part of the fpecific charaĉter throughout this genus. Both are certainly miftaken in affigning thirty feet to each fide of this creature. Degeer mentions thirty-one ; and in an unmutilated fpecimen we have, two legs may be perceived at every joint except thofe neareft the head. The body is flat, the fhields flightly fcabrous, and the antennæ clavated, the laft we deem more charaĉteriftic than the number of the feet.

This very curious creature is local, being rare in moft places.

F 3                                    PLATE

# 345

# PLATE CCCXLV.

## FIG. I.

### PHALÆNA GRANDIS.

#### Grey Arches Moth.

#### *GENERIC CHARACTER.*

Antennæ taper from the base. Wings in general deflexed when at rest. Fly by night.

#### *SPECIFIC CHARACTER.*

Wings whitish, variegated with black, and waved or arched transverse streaks. A large eye-shaped spot in the middle, and a black character in the posterior angle.

Phalæna Grandis: alis albicantibus nigro-varie undatis: stigmatibus magnis subocellaribus, litura prope anglum posticum nigrum.

―――――――――――

An Insect well known amongst English collectors by the name of Grey Arches Moth, from the characteristic arched double lines across the superior wings. It is altogether unnoticed by Linnæus or Fabricius, though figured by Sepp. vol. ii. *tab.* 27. It is esteemed a scarce species in this country.

F 4                                                        F I G,

F I G. II. III.

PHAL ÆNA   S P I N U L A.

*S P E C I F I C   C H A R A C T E R.*

Wings variegated brown and grey, with obfcure tranfverfe bars.
Three diftinct black pointed characters near the apex.

PHALÆNA SPINULA: alis fufco cinereis maculis ftrigifque ob-
    fcuris lituris tribus acutiufculis diftinctis nigris
    ad apicem.

———————

The fpecimens, *fig.* 2 and 3, appear at the firft view two very
diftinct Infects; yet on the moft attentive comparifon of the characteriftic marks, we are inclined to confider them as the two fexes
of the fame fpecies, notwithftanding the diffimilarity of their colours in general. The kind reprefented at *fig.* 2, and which from
its fetaceous antennæ is evidently the female, has been placed in
Englifh cabinets as a fpecies fomewhat analogous to the Phalæna
Exoleta, or Sword-blade Moth, under the trivial appellation of the
*fcarce* Sword-blade Moth. The other, which from the pectinated
ftructure of its antennæ, is obvioufly the male, is equally uncommon.

 We have obferved feveral figures of this fpecies different only in
colour in the works of Ernft, and one in particular nearly corre-
fponding with that reprefented in the annexed plate at *fig.* 2, which
he calls *fpinula*; a name we have ventured to adopt, as the Infect
is neither defcribed by Linnæus nor Fabricius.

P L A T E

# 346

# PLATE CCCXLVI.

## FIG. I.

### MUSCA GROSSA.

GREAT BLACK FLY.

DIPTERA.

*GENERIC CHARACTER.*

A foft flexible trunk, with lateral lips at the end. No palpi.

*SPECIFIC CHARACTER*

AND

*SYNONYMS.*

Body hairy, black. Wings ferruginous at the bafe.

MUSCA GROSSA: pilofa nigra, alis bafi ferrugineis. *Linn. à Gmel.*
    *T.* 1. *p.* 5. *p.* 2845. *fp.* 75.
    *Fn. Suec.* 1837.
    *Fab. Sp. Inf.* 2. *p.* 441. *n.* 30.
    *Schæff. Icon. tab.* 108. *fig.* 6.
    *Degeer. Inf.* 6. *p.* 21. *n.* 1. *tab.* 1. *fig.* 1.

---

  The largeft of the Mufca genus found in this country. Breeds in dung.

FIG.

## FIG. II.

## MUSCA BICINCTA.

### Double Belted Fly.

### *GENERIC CHARACTER.*

Black. Sides of the thorax and two belts acrofs the abdomen yellow.

Musca Bicincta: nigra, antennis elongatis, thorace lateribus
punctis abdomineque cingulis duobus flavis.
*Linn. a Gmel. T. 1. p. 5. p. 2872. fp. 38.*
*Fab. Sp. Inf. 2. p. 427. n. 30.*
*Degeer. Inf. 6. p. 126. n. 16. t. 7. fig. 16.*

---

## FIG. III.

## MUSCA VIBRANS.

### Vibratory Fly.

### *SPECIFIC CHARACTER*

### AND

### *SYNONYMS.*

Wings tranfparent, black at the tip. Head red.

Musca

# PLATE CCCXLVI.

55

MUSCA VIBRANS: alis hyalinis apice nigris capite rubro.
Linn. a Gmel. T. 1. p. 5. p. 2855. sp. 112.
Fab. Sp. Inf. 2. p. 450. n. 81.
Degeer. Inf. 6. p. 32. n. 11. t. 1. fig. 19.
Geoffr. Inf. p. 2. p. 494. n. 4.

Remarkable for the continual vibratory motion of its wings.

## FIG. IV.

## MUSCA NOCTILUCA.

*SPECIFIC CHARACTER*

AND

*SYNONYMS.*

Somewhat hairy, black ; two pellucid spots on the first segment of the abdomen.

MUSCA NOCTILUCA: subtomentosa atra, abdominis segmento
primo maculis duabus pellucidis. *Linn. a Gmel.*
T. 1. p. 5. p. 2874. sp. 48.
Faun. Suec. 1814.
Fab. Sp. Inf. 2. p. 431. n. 54.

FIG.

## FIG. V.

## MUSCA SCYBALARIA.

### *SPECIFIC CHARACTER.*

Reddish brown, an obscure dot in the Wings.

MUSCA SCYBALARIA: rufa ferruginea, alis puncto obscuriore.
      *Linn. a Gmel. T.* 1. *p.* 5. *p.* 2853. *sp.* 104.
      *Faun. Suec.* 1860.
      *Fab. Sp. Inf.* 2. *p.* 449. *n.* 72.
      *Scop. Carn.* 896.

Found on Dung.

      PLATE

# 347

# PLATE CCCXLVII.

## FIG. I.

### PHALÆNA APRILINA,
SCARCE MERVEILLE DU JOUR MOTH.

### *GENERIC CHARACTER.*

Antennæ taper from the bafe. Wings in general deflexed when at reft. Fly by Night.

### *SPECIFIC CHARACTER*
### AND
### *SYNONYMS.*

Thorax crefted. Wings deflexed, green: a black mark and tranfverfe band; and a fingle row of black triangular dots near the apex.

PHALÆNA APRILINA: criftata alis deflexis viridibus: macula fafciaque atris apice punctorum trigonum ferie unica. *Fab. Ent. Syft. T.* 3. *p.* 2. *p.* 103. *fp.* 306.

PHALÆNA runica *Linn.*

---

Linnæus has made fome confufion between the two fpecies of Phalæna Aprilina and runica in feveral of his works. In the laft edition by Gmelin, our Infect ftands as the P. *runica*; and in the Entomologia Syftematica of Fabricius, which we have in this inftance preferred, it is the Phalæna Aprilina.

The

The Englifh Entomologift is indebted to the affiduity of the late Duchefs of Portland for the difcovery of this extremely rare fpecies in England. It feeds on the Oak.

---

## FIG. II.

### PHALÆNA PINASTRI.

*SPECIFIC CHARACTER*

AND

*SYNONYMS.*

Thorax crefted. Wings deflexed, blackifh: oblique broad fpace along the exterior margin grey.

PHALÆNA PINASTRI: criftata alis deflexis nigris : margine tenuiori anguloque ani obfcure cinereis. *Linn. Syft. Nat.* 2. 851. 160. *Fab. Ent. Syft. T. 3. p. 2. p.* 101. *fp.* 302.

---

Not fo rare as the preceding but ftill much efteemed by the Englifh Entomologift.—Feeds on the Pine.

# FIG. III.

## PHALÆNA GEMINA.

### SPECIFIC CHARACTER.

First wings greyish brown, with two transverse broad bands and two connected white spots, and a minute dot in the middle.

PHALÆNA GEMINA: spirilinguis criftata, alis fuperioribus cinereo-fufcentibus, fafciis duabus ftrigofis maculifque duabus niveis intermediis. *Beckwith's paper tranf. Linn. Soc. Vol.* 2. *p.* 4.

––––––––––

The larva is of a pale yellow with a red head. It feeds on the Poplar, and about the beginning of October enclofes itfelf between two leaves, which it unites at the edges by means of many ftrong threads, and becomes a pupa. The Moth burfts forth about the end of May or beginning of June.

PLATE

# 348

# PLATE CCCXLVIII.

## FIG. I.

## CURCULIO LATIROSTRIS.

### *GENERIC CHARACTER.*

Antennæ fubclavated and feated in a roftrum or probofcis, which is of a horny fubftance and prominent.

### *SPECIFIC CHARACTER*

### AND

### *SYNONYMS.*

Snout fhort, broad, and flattifh. Wing-cafes brown, with two black fpots: apex white.

CURCULIO LATIROSTRIS: roftro latiffimo plano, elytris apice albis: punctis duobus nigris. *Linn. a Gmel.* *T.* 1. *p.* 4. *p.* 1783. *fp.* 36c.

G

FIG.

# FIG. II.

## CURCULIO PARAPLECTICUS.

### SPECIFIC CHARACTER

#### AND

### SYNONYMS.

Cylindrical, yellowish brown. Wing-cafes terminated in an acute point.

CURCULIO PARAPLECTICUS: cylindricus fubcinereus, elytris mucronatis. *Fn. Sv.* 604.—*Linn. a Gmel. T.* 1. *p.* 4. *p.* 1750. *fp.* 34.—*Schœff. Icon. t.* 44. *fig.* 1.

---

# FIG. III.

## CURCULIO ALBINUS.

### SPECIFIC CHARACTER.

Black, front of the head and tip of the wing-cafes white. Thorax tuberculated.

CURCULIO ALBINUS: niger, fronte anoque albis, thorace tuberculato. *Gmel. a Linn. T.* 1. *p.* 4. *p.* 1783. *fp.* 79.

---

This, as well as the two preceding fpecies is very rare.

PLATE

349

# PLATE CCCXLIX.

## FIG. I.

## PHALÆNA DOLABRARIA.

### Scorched Wing Moth.

### *GENERIC CHARACTER.*

Antennæ taper from the bafe. Wings in general deflexed when at reft. Fly by night.

### *SPECIFIC CHARACTER*

### AND

### *SYNONYMS.*

Wings yellow, with numerous ferruginous tranfverfe ftreaks. Anal angle violet.

Phalæna Dolabraria : alis flavis : ftrigis ferrugineis anguloque ani violaceo. *Linn. Syft. Nat. T.* 1. *p.* 4.
*p.* 2451. *Fab. fp. Inf.* 2. *p.* 245. *n.* 21.
*fp.* 207.

G 2  FIG.

## F I G. II.

## PHALÆNA URTICATA.

*SPECIFIC CHARACTER*

AND

*SYNONYMS.*

Antennæ like a briſtle. Wings white, with bands of brown ſpots. Thorax and tail yellow.

PHALÆNA URTICA : ſeticornis alis albis fuſco faſciato-maculatis, thorace anoque flavis. *Linn. Syſt. Nat.* 2. 873. 272.
　　　　*Fab. Ent. Syſt. T.* 3. *p.* 2. *p.* 209. *ſp.* 299.
　　　　*Roeſ. Inſ.* 1. *phal.* 4. *tab.* 14.
　　　　*Degeer. Inſ.* 1. *tab.* 28. *fig.* 18. 19.
　　　　*Geoffr. Inſ.* 2. 135.

The larva conceals itſelf in a kind of cylinder, which it forms by rolling up the edges of the nettle leaves on which it feeds. It is whitiſh, with a dark dorſal line, head black, and two ſpots of the ſame colour on the ſegment next the head.

FIG.

PLATE CCCXLIX. 65

## FIG. III.

## PHALÆNA LYNCEATA.

### SPECIFIC CHARACTER.

Wings white, with two brown tranfverfe bands, and a brown fpot near the apex.

PHALÆNA LYNCEATA : alis albis : fafciis duabus punctoque apicis fufcis. *Fab. fpec. Inf. 2. p. 262. n. 129. Gmel. Linn. Syft.* 2478.

---

This Infect, though very common in our woods, was unknown to Fabricius before his vifit to Great Britain; he firft defcribed it in the fpecies Infectorum, under the fpecific name of Lynceata, as an Englifh Infect : it has fince appeared in his other publications, and has been inferted by Gmelin in the laft Edition of the Syftema Naturæ.

The Linnean defcription of P. ocellata coincides fo nearly with this Infect, that we may doubt the propriety of feparating them; they are probably varieties only of the fame Species.

It is very common in June.

G 3 PLATE

# 350

# PLATE CCCL.

## FIG. I.

### PHALÆNA DROMEDARIUS.

IRON PROMINENT MOTH.

*GENERIC CHARACTER.*

Antennæ taper from the bafe. Wings in general deflexed when at reft. Fly by night.

*SPECIFIC CHARACTER*

AND

*SYNONYMS.*

Wings deflexed, clouded, a large tufted dentation at the pofterior margin : bafe yellowifh.

PHALÆNA DROMEDARIUS : alis deflexis : anticis nebulofis dorfo dentatis : litura bafeos anique flavefcentibus. *Linn. Syft. Nat.* 2. 827. 62.—*Fab. Ent. Syft.* T. 3. *p.* 1. *p.* 444. *fp.* 113. Ammiral. *Inf. tab.* 14.

## FIG. II.

### PHALÆNA CHRYSOGLOSSA.

*SPECIFIC CHARACTER*

AND

*SYNONYMS.*

Thorax crefted. Firft wings fomewhat falcated or hooked, greyifh, with three ftreaks, and two kidney-fhaped fpots in the middle.

G 4

PHALÆNA

PHALÆNA CHRYSOGLOSSA : fpirilinguis criftata, alis fuperioribus
grifeis fubfalcatis ftrigis tribus albis primoribus
abbreviatis. *Linn. Tranf. Vol.* 2. 1. *p.* 6.

---

One of the rare fpecies of Phalænæ, defcribed by the late Mr.
Beckwith in the Linnæan tranfactions. The larva is remarkably
flender, and of a green colour ; it was found upon the fallow near
Brent-Wood on the 18th of June, went into the earth about a week
after, and the Moth was produced on the 24th of July.

---

## F I G.  III.

## PHALÆNA RUBRICOLLIS.

### RED-NECKED MOTH.

### *SPECIFIC CHARACTER.*

Blackifh, collar crimfon ; end of the abdomen yellow.

PHALÆNA RUBRICOLLIS : atra, collari fanguineo, abdomine flavo.
*Linn. a Gmel. T.* 1. *p.* 4. *p.* 2446. *fp.* 113.
*Schæff. Icon. t.* 59. *f.* 8. 9.

---

This fingular creature was found in Coombe Wood in the month
of June. The larva is hairy, dark, ftriped with black, and has a
white triangular mark on the head. It feeds on the pine,
beech, &c.

P L A T E

# 351

1

2

# PLATE CCCLI.

## FIG. I.

## CICINDELA SYLVATICA,

### *GENERIC CHARACTER.*

Antennæ fetaceous. Maxillæ or jaws advanced confiderably be-fore the head. Eyes prominent. Thorax roundifh and margined.

### *SPECIFIC CHARACTER.*

Black, a white waved band, and two dots of the fame colour on the Wing-cafes.

CICINDELA SYLVATICA : nigra, elytris fafcia undata punctifque duobus albis. *Linn. Gmel. T.* 1. *p.* 4. *p.* 1922. *fp.* 8.

Cicindela atra, coleopteris maculis fex albida fafciaque albis. *Faun. Suec.* 1. *n.* 549.

Cicindela fupra nigra, fubtus viridis nitida, &c. *Degeer. Inf.* 4. *p.* 114. *t.* 4. *f.* 7.

———

A very fcarce Englifh Infect.

F I G.

## F I G. II.

## C I C I N D E L A   A Q U A T I C A.

*SPECIFIC CHARACTER.*

Shining, bronzed, head ſtriated.

Cicindela Aquatica : ænea nitida, capite ſtriato.   *Linn. a Gmel.*
*T.* 1. *p.* 4. *p.* 1925. *ſp.* 14.—*Fn. Sv.* 752.

Cicindela Pusilla.   *Schreb. Inſ.* 6.

Bupreſtis fuſco-æneus.   *Geoff. Inſ. p.* 1. *p.* 157. *n.* 31.

---

Extremely common in ſome moiſt ſituations.

PLATE

# 352

# PLATE CCCLII.

## FIG. I.

### PHALÆNA DIVES.

#### Brocade Moth.

#### *GENERIC CHARACTER.*

Antennæ taper from the bafe. Wings in general deflexed when at reft. Fly by Night.

#### *SPECIFIC CHARACTER*

##### AND

#### *SYNONYMS.*

Wings brown: bafe, central fpots and broad tranfverfe bar near the exterior end grey; a black line at the bafe, a bidentated dark line along the apex, and a black mark near the pofterior margin.

PHALÆNA DIVES: alis fufcis: bafi ftigmatibus fafciaque poftica bidentata cinereis, linea bafeos alteraque pofteriori nigris.

An undefcribed fpecies, known by the Englifh name of Brocade Moth.

## FIG. II.

### PHALÆNA TRIMACULA.

#### *SPECIFIC CHARACTER.*

Wings cinereous clouded with brown; bafe, apex and a broad tranfverfe bar acrofs the middle white.

PHALÆNA

PHALÆNA TRIMACULA: alis cinereis nigro-nebulofis: bafi pal-
lidiore, fafcia lata apiceque albis.

---

This feems to be no other than the Bombyx trimacula of the
Vienna catalogue, *Wien. Verz.* 59. *No.* 4. and the B. trifafcia of
Efper, *p.* 242. *t.* 46. *fig.* 1—2; a fpecies unnoticed in the Ento-
mologia Syftematica of Fabricius.

This is one of the rare Infects difcovered by the late Duchefs of
Portland.

---

## F I G. III.

## PHALÆNA FLAVICORNIS.

### YELLOW-HORNED PHALÆNA.

### *SPECIFIC CHARACTER*

### A N D

### *SYNONYMS.*

Firft wings greyifh tinged with yellow, and marked tranfverfely
with three black ftreaks. Antennæ yellow.

PHALÆNA FLAVICORNIS: alis primoribus cinereis: ftrigis tribus
atris, antennis luteis. *Fn. Sv.* 1204.—*Linn.
Syft. Nat. a Gmel. T.* 1. *p.* 5. *p.* 2575. *fp.* 182.
—*Fab. fpec. Inf.* 2. *p.* 238. *n.* 140.

---

A fcarce Moth, faid to feed on fruit-trees.—Cabinet of
A. M'Leay, Efq.

PLATE

353

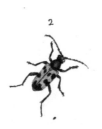

# PLATE CCCLIII.

## FIG. I.

## LEPTURA NIGRO-LINEATA.

### BLACK-STREAKED WASP BEETLE.

### *GENERIC CHARACTER.*

Antennæ gradually tapering towards the end. Elytra narrower towards the extremity. Thorax round and slender.

### *SPECIFIC CHARACTER.*

Black, with a metallic tinge. Wing-cafes yellow, with three interrupted longitudinal lines of black.

LEPTURA NIGRO-LINEATA : nigro-ænea elytris flavis : bafi lineifque tribus longitudinalibus interruptis nigris. *Marfh. Mfs.*

---

A very rare and curious fpecies.—In the collection of Mr. Francillon.

FIG.

## F I G.  II.

## LEPTURA  SEX-MACULATA.

Six Spot Wasp Beetle.

### SPECIFIC CHARACTER.

Black. Wing-cafes yellowifh, with three black fpots connected
with the outer margin on each.

Leptura Sex-maculata : nigra, coleoptris teftaceis: maculis
fex nigris margine connexus. *Gmel. a Linn.*
*T.* i. *p.* 4. *p.* 1871. *fp.* 11.
*Fab. Sp. Inf.* i. *p.* 248. *n.* 16.

Leptura teftaceo maculata.—*Degeer. Inf.* 5. *p.* 133. *n.* 9.

———————————

Uncommon in Great Britain. In our fpecimen, the anterior
black mark on the wing-cafes is interrupted, and appears like two
diftinct fpots.

P L A T E

354

# PLATE CCCLIV.

## FIG. I.

### PHALÆNA RUNICA.

COMMON MERVEILLE DU JOUR MOTH.

### GENERIC CHARACTER.

Antennæ taper from the base. Wings in general deflexed when at rest. Fly by Night.

### SPECIFIC CHARACTER

AND

### SYNONYMS.

Thorax crested. First wings greenish with black marks; and a row of triangular spots behind.

PHALÆNA RUNICA : cristata alis deflexis : anticis virescentibus, maculis variis atris, postice utrinque punctis trigonis. *Fab. Ent. Syst. T.* 3. *p.* 2. *p.* 102. *sp.* 305.

PHALÆNA APRILINA. *Linn. Syst. Nat. a Gmel. p.* 2561. *sp.* 1123.

The

The two black femicircles on the thorax and double row of triangular fpots at the ends of the pofterior wings, are mentioned as peculiar characters of this fpecies ; the latter is, however, liable to confiderable variations, the fpots being in general, crowded in a confufed feries, and forming an irregularly interrupted line. It may be eafily diftinguifhed from the Phalæna Aprilina by its fuperior fize, the colours are lefs vivid, and it is deftitute of the broad tranfverfe bar, which is confpicuous on the upper wings of Phalæna Aprilina.

The larva is fmooth, of an afh colour with fquare brownifh fpots. It feeds on the Oak.

## FIG. II.

### PHALÆNA LUSORIA.

#### SPECIFIC CHARACTER

AND

#### SYNONYMS.

Thorax crefted. Wings incumbent, greyifh ; a black lunar mark, with a fmall approximate point in the middle of the firft pair, and a triangular fpot of the fame colour on the anterior part of the thorax.

PHALÆNA LUSORIA : criftata alis incumbentibus glaucis : lunula thoraceque antice atris. *Linn. Syft. Nat.* 2. 831. 74.—*Fab. Ent. Syft. T.* 3. *p.* 2. *p.* 64. *fp.* 179.

Phalæna Luforia : alis primoribus glaucis : lunula thoraceque anterius ferrugineis. *Gmel. Linn. Syft. Nat. T.* 1. *p.* 5. *p.* 2441. *fp.* 74.

From the collection of Mr. Francillon.

PLATE

# 355

# PLATE CCCLV.

## FIG. I.

### PHALÆNA FRANCILLANA.

#### GENERIC CHARACTER.

Antennæ taper from the bafe. Wings in general deflexed when at reft. Fly by night.

#### SPECIFIC CHARACTER.

Pale yellow, with two oblique brown ftreaks acrofs each of the fuperior wings.

PHALÆNA FRANCILLANA : alis pallide flavis : ftrigis duabus brunneis. *Fab. Ent. Syft. T.* 3. *p.* 2. *p.* 264. *fp.* 94.

---

Fabricius, when in England, named this elegant little nondefcript Francillana, in compliment to Mr. Francillon, whofe exquifite collection of Infects afforded him fuch ample affiftance in completing his fpecies Infectorum and Entomologia Syftematica, and we are happy

H                                                                in

in the opportunity to perpetuate the name, as a fmall acknowledg-
ment for the confidential freedom with which its worthy poffeffor has
at all times permitted us to defcribe and copy whatever our own Cabinet was deficient in.—The fpecimen is in the collection of Mr.
Francillon.

## F I G.  II.

## PHALÆNA LITTERANA.

### *SPECIFIC CHARACTER.*

Wings rhombic: firft pair green, with little elevated tufts and
black characters.

PHALÆNA LITTERANA: alis rhombeis: anticis viridibus: charac-
teribus atris.  *Linn. Syft. Nat.* 2. 876. 288.—
*Fab. Ent. Syft. T.* 3. *p.* 2. *fp.* 271.

Not unlike Phalæna Squamana, figured in a former part of this
Work, except in the black characteriftic marks on the fuperior
wings.—This is a very rare and beautiful little creature.

F I G.

PLATE CCCLV.                79

# FIG. III.

## PHALÆNA CYNOSBANA.

### SPECIFIC CHARACTER.

Anterior wings dark brown, white at the tips.

PHALÆNA CYNOSBANA : alis anticis fuscis apice albis.   *Fab.*

PYRALIS CYNOSBANA.   *Fab. Ent. Syst.* 3. *b.* 283. 167.

Tinea Cynosbatella.   *Linn. Syst. Nat.* 887. 368.—*Fn. Sv.* 1397.

---

Linnæus, and after him Fabricius, refers to the works of De Geer and Merian for figures of this Insect, neither of which are in our opinion sufficient authorities; that of De Geer is in particular miserable, and so far from leading to ascertain the species, may mislead us as to the identity of its genera.   Some years since, this Insect was deemed a nondescript by English collectors, and named in compliment to a much respected *Aurelian* Beckwithiana ; but agreeing in every respect with the Linnæan *Cynosbana,* we must unavoidably reject that name, though generally adopted.

It is very common, and according to Linnæus breeds in the buds of roses.

H 2                    FIG.

# FIG. IV.

## PHALÆNA EVONYMELLA.

### WHITE ERMINE MOTH.

### SPECIFIC CHARACTER.

Wings white, with about fifty black points.

Tinea, with white upper wings and black points; inferior wings
   brown. Tinea alis fuperioribus albis: punctis
   nigris, inferioribus fufcis. *Geoff. Inf.* 2. 183. 4.

PHALÆNA EVONYMELLA : alis anticis niveis : punctis 50 nigris.
   *Linn. Syft. Nat.* 2. 885. 350.
   *Fn. Sv.* 1363.
   *Fab. Ent. Syft. T.* 3. *p.* 2. 289. 12.

———————————

Linnæus confidered the two fpecies of Ermine Moths, Padella
and Evonymella, fufficiently characterifed by the number of black
fpots on the fuperior wings. To the firft he affigns twenty, and
to the latter fifty on each wing.

Thofe who have attended particularly to thofe fpecies are aware,
that the number of fpots are by no means conftant, and may have
frequently obferved even more fpots on one wing than the other in
the fame individual, as occurs in the fpecimen we have figured. This
has occafioned fome confufion between the two fpecies, and we expreffed fome doubts refpecting them in the defcription of one of the
             earlieft

# PLATE CCCLV.

81

earlieſt plates in this work. Since that time we have had more opportunities of aſcertaining the two kinds, and have no heſitation in
admitting them as two ſpecies. Not that the variation in the number
of ſpots will allow us to admit the definition of Linnæus uncondi-
tionally. It may be ſaid that thoſe on the P. Padella are about
twenty, and thoſe on the P. Evonymella ſometimes amount to fifty
or more, and the colour of the ſuperior wings is lighter in the
latter than the former.

In admitting this, the name Evonymella adopted in the ninth plate,
muſt be changed to Padella; and the Inſect before us be conſidered
as the true Evonymella.

## FIG. V.

## PHALÆNA FALCATELLA.

### TRIANGLE-MARKED LIGHT HOOK-TIP.

### SPECIFIC CHARACTER.

Wings hooked at the apex, whitiſh, decuſſated with obſcure bars,
and a large triangular brown ſpot on the poſterior margin.

PHALÆNA FALCATELLA: alis falcatis albis: faſciis obſcuris de-
cuſſatis maculaque magna trigona marginis tenui-
oris fuſca.

Very ſcarce and undeſcribed.

H 3  PLATE

# 356

# PLATE CCCLVI.

## ELATER FERRUGINEUS.

### COLEOPTERA.

### GENERIC CHARACTER.

Antennæ filiform. Palpi four. An elaſtic ſpine at the extremity of the thorax on the under ſide, by means of which it ſprings up when placed on the back.

### SPECIFIC CHARACTER.

Thorax and wing-caſes ferruginous. Body and poſterior margin of the thorax black.

ELATER FERRUGINEUS: thorace elytriſque ferrugineis, corpore thoraceque margine poſteriore nigris. *Linn. a Gmel. T.* 1. *p.* 4. *p.* 1906. *ſp.* 20.

---

Scarce. From the collection of Mr. Francillon.

H 4                                                    FIG.

## FIG. II.

## ELATER PECTINICORNIS.

### *SPECIFIC CHARACTER.*

Above greenifh with a dull metallic glofs.    Antennæ of the male large and pectinated.

ELATER PECTINICORNIS: thorace elyftrifque æneis, antennis
maris pectinatis. *Linn. Gmel. T.* 1. *p.* 4. *p.* 1909.
*fp.* 32.

———————

The elegant antennæ of this Infect is very characteriftic.    It is not an uncommon fpecies.

PLATE

# 357

# PLATE CCCLVII.

## FIG. I.

### PHALÆNA BENTLEIANA.

#### *GENERIC CHARACTER.*

Antennæ taper from the bafe. Wings in general deflexed when at reft. Fly by Night.

#### *SPECIFIC CHARACTER.*

Reddifh brown, with numerous ftreaks of a filvery yellow.

PHALÆNA BENTLEIANA : alis fufco ferrugineis : ftrigis punctifque numerofis argenteo-flavis.

———————

As no Infect has yet appeared to record the memory of that indefatigable collector of Englifh Infects, Mr. Bentley, we are induced to affign his name to this beautiful and hitherto namelefs Species— It is extremely fcarce.

## FIG. II.

### PHALÆNA ILICANA.

#### *SPECIFIC CHARACTER*

Anterior wings greyifh brown, with brown fpots, a folitary black fpot in the middle.

PHALÆNA

PHALÆNA ILICANA: alis anticis fufco-cinereis: punctis fufcis; centrali folitario atro.  *Fab. Ent. Syft.* 3. *b.* 266. 100.

----

Fabricius defcribes this as an Englifh Infect.  It is rare, and not hitherto figured.

----

## FIG. III.

## PHALÆNA BIFASCIANA.

### *SPECIFIC CHARACTER.*

Anterior wings teftaceous, with two whitifh bands, and four diftinct undulated ftreaks and fpots of black.

PHALÆNA BIFASCIANA: alis anticis teftaceis, fafciis duabus albidis, ftrigis quatuor undatis maculifque nigris.

----

A very uncommon Infect, and not noticed by any author.

FIG.

PLATE CCCLVII. 87

# FIG. IV.

## PHALÆNA LEEANA.

### SPECIFIC CHARACTER

#### AND

### SYNONYMS.

Wings pale, yellowiſh, with a brown ſpot in the middle.

PHALÆNA LEEANA : alis pallidis: macula centrali fuſca.
*Gmel. Linn. Syſt. Nat. p.* 2497.
*Fab. Spec. Inſect.* 2. *p.* 276. *n.* 2.

Taken in June and July.  Not uncommon.

PLATE

# 358

1

2

# PLATE CCCLVIII.

## FIG. I.

## NECYDALIS HUMERALIS.

### GENERIC CHARACTER.

Antennæ fetaceous or filiform. Wing-cafes lefs than the wings, and either narrower or fhorter than the abdomen. Tail fimple.

### SPECIFIC CHARACTER

### AND

### SYNONYMS.

Wing-cafes narrow, and tapering to a point, black, yellow at the bafe.

NECYDALIS HUMERALIS: elytris nigris bafi flavis.—*Fab. fpec. Inf.* 1. *p.* 263. *fp.* 5.

NECYDALIS HUMERALIS.—*Gmel. T.* 1. *p.* 4. *p.* 1880. *fp.* 18.

Necydalis (muralis) elytris fubulatis fufca, humeris flavis, pedibus fimplicibus.—*Forft. nov. inf. fp.* 1. *p.* 48. *n.* 48.

---

A fcarce fpecies, defcribed by Fabricius and Gmelin as a native of this country.

FIG.

# F I G. II.

## NECYDALIS SIMPLEX.

*SPECIFIC CHARACTER*

AND

*SYNONYMS.*

Wings teftaceous.   Legs fimple.

NECYDALIS SIMPLEX: elytris teftaceis, pedibus fimplicibus.—
*Fab. fpec. Inf.* 1. *p.* 264. *fp.* 9.
*Gmel. T.* 1. *p.* 4. 1881. 10.

Cantharis phthyfica: *Scop. Ent. Carn.* 144.

———————

Supppofed by Gmelin to be a variety of Necydalis podagrariæ.

P L A T E.

# 359

# PLATE CCCLIX.

## PHALÆNA TRITOPHUS.

### Aspen Prominent Moth.

### GENERIC CHARACTER.

Antennæ taper from the bafe. Wings in general deflexed when at reft. Fly by Night.

### SPECIFIC CHARACTER.

Wings deflexed, a prominent tuft or tooth on the pofterior margin, brown, clouded; in the middle, a white ring, enclofing a ferruginous lunar mark.

Phalæna Tritophus : alis deflexis dorfo dentatis fufco nebulofis : lunula media ferruginea alba cincta. *Fab. Ent. Syft. T.* 3. *p.* 1. *p.* 448. *fp.* 108.

---

The larva of this fine Infect is green, with a brown head, obtufe tail, and three elevations or gibbofities on the back. It feeds on the *Populo tremulo,* from whence we have deduced its Englifh name of Afpen Prominent Moth.

Phalæna tritophus is extremely fcarce in this country.

PLATE

# 360

# PLATE CCCLX.

## FIG. I.

### PHALÆNA ERICÆ.

TRANSVERSE-STREAK HEATH MOTH.

### GENERIC CHARACTER.

Antennæ taper from the bafe. Wings in general deflexed when at reft. Fly by night.

### SPECIFIC CHARACTER.

Anterior Wings brown, with two undulated ftreaks, and fpots of white. Pofterior wings pale.

PHALÆNÆ ERICÆ: alis anticis fufcis: ftrigis duabus undatis maculis ordinariis lineolifque albis, pofticis pallidis.

---

A non-defcript fpecies of the noctua family, and very rare. Found on heaths.

I

FIG.

## FIG. II.

## PHALÆNA LINEOLA.

### SHORT-LINE MOTH.

### *SPECIFIC CHARACTER.*

Anterior wings ferruginous grey, with undulated ftreaks. A fmall oblique line in the middle, and a row of brown points along the exterior margin.

PHALÆNA LINEOLA : alis anticis grifeo-ferrugineis: ftrigis undatis lineola obliqua in medio punctifque poftice fufcis.

More frequent than the preceding fpecies, and feems to be figured in the works of Ernft and Efper, but certainly not defcribed by any fyftematic author.

The colour varies in different fpecimens from ferruginous to greyifh or livid colour.

FIG.

# PLATE CCCLX.                95

## F I G. III.

## PHALÆNA MAPPA.

MAP-WING SWIFT MOTH.

### SPECIFIC CHARACTER.

Wings brown, with large irregular waved marks of a livid colour, and four diftinct white triangular fpots near the apex.

PHALÆNA MAPPA : alis fufcis : lituris magnis irregularibus lividis punctifque quatuor poftice albidis.

---

A Moth of the *Hepialus*, a new genus in the Fabrician Syftem, including only ten fpecies, neither of which agrees with our Infect, and we apprehend it is not defcribed by any other Author.

In the Linnæan Syftem, this can only be regarded as a family of the Phalæna tribe. The Englifh collectors have denominated this family Swifts, as noticed already in the defcriptions of Humuli and Hecta, (plate 274. fig. 1, 2, 3.)—The prefent Infect may be confidered as one of the rareft Englifh undefcribed fpecies, and the many windings of the numerous marks on the anterior wings, immediately fuggeft the appropriate fpecific name of Mappa.

# LINNÆAN INDEX

TO

## VOL. X.

### COLEOPTERA.

| | | | | Plate | Fig. |
|---|---|---|---|---|---|
| Ptinus pectinicornis | - | - | - | 326 | |
| Chryſomela marginella | - | - | - | 335 | |
| Curculio albinus | - | - | - | 348 | 3. |
| —— paraplecticus | - | - | - | 348 | 2. |
| —— latiroſtris | - | - | - | 348 | 1. |
| Leptura ſex maculata | - | - | - | 353 | 2. |
| —— nigro-lineata | - | - | - | 353 | 1. |
| Necydalis ſimplex | - | - | - | 358 | 2. |
| —— humeralis | - | - | - | 358 | 1. |
| Elater ferrugineus | - | - | - | 356 | 1. |
| —— pectinicornis | - | - | - | 356 | 2. |
| Cicindela ſylvatica | - | - | - | 351 | 1. |
| —— aquatica | - | - | - | 351 | 2. |

### HEMIPTERA.

| | | | | | |
|---|---|---|---|---|---|
| Blatta Lapponica | - | - | - | - | 352 |
| —— Germanica | - | - | - | - | 341 |

K

LEPI.

# I N D E X.

## L E P I D O P T E R A.

———— con-

# INDEX.

## NEUROPTERA.

## HYMENOPTERA.

K 2                                        DIPTERA.

# INDEX.

## DIPTERA.

## APTERA.

ALPHA-

# ALPHABETICAL INDEX

TO

# VOL. X.

# INDEX.

# INDEX.

F I N I S.

Printed by Bye and Law, St. John's-Square, Clerkenwell.

THE

NATURAL HISTORY

OF

# BRITISH INSECTS;

EXPLAINING THEM

IN THEIR SEVERAL STATES,

WITH THE PERIODS OF THEIR TRANSFORMATIONS,
THEIR FOOD, ŒCONOMY, &c.

TOGETHER WITH THE

HISTORY OF SUCH MINUTE INSECTS

AS REQUIRE INVESTIGATION BY THE MICROSCOPE.

THE WHOLE ILLUSTRATED BY

COLOURED FIGURES,

DESIGNED AND EXECUTED FROM LIVING SPECIMENS.

By E. DONOVAN, F.L.S.

VOL. XI.

LONDON:

PRINTED FOR THE AUTHOR,

And for F. C. and J. RIVINGTON, Nº 62, ST. PAUL'S CHURCH-YARD.

MDCCCVI.

# ADVERTISEMENT.

AS a general hiftory of the Entomological productions of Great Britain, this Work has been long acknowledged the moft copious hitherto fubmitted to the Public. The firft part, comprifing no lefs than ten volumes, having already appeared in monthly numbers, the author does not conceive it in any refpect incumbent on him to enlarge on its pretenfions to notice. Whatever may be its merits, the author muft, in candour, allow they have been amply appreciated by the liberality of that public, who, for the fpace of ten years, were pleafed to fanction it with the beft teftimony of their approbation. The publication of this work has been fome time difcontinued, but the occafion of this delay is fufficiently known. The Author had then fulfilled his firft engagements, fo far as related to this work, and was unwilling to trefpafs beyond the limits thofe engagements prefcribed.—He ftated, notwithftanding, at the conclufion of the work, that no confideration fhould permit him to entirely abandon his entomological purfuits: that his attention would be directed to a fcience in which the paft indulgence of the public had induced him to believe he might yet be ufeful; and that, fhould a number of new and valuable infects occur, they would be certainly added in a fupplemental form to improve the former work. Since that period, the author has been led to conceive, his endeavours might not prove unacceptable in elucidating

elucidating the fcience of Britifh Entomology, upon a ftill more extenfive fcale than even this fuggeftion intimated; and it is under this idea, he once more folicits the attention of his former fubfcribers and the public, in favour of a fecond part of his original undertaking.

The object of the ten preceding volumes was avowedly to comprehend a felection only of the moft beautiful, or otherwife particularly interefting fpecies of infects from all the various claffes, but more efpecially from that of the Lepidopterous tribes; more was not promifed, and could not have been expected. When, therefore, the author ventures to extend the limits of his original defign, the motives for it fhould be unequivocally ftated. It is not upon the addition merely of a few felect fubjects omitted in the former work, either in a fupplemental, or any other form, that he now conceives he ought to reft his claims to further notice, but by declaring what it is his intention to fulfil; that the future volumes, with the preceding, fhall comprife, collectively, a general hiftory, and elucidation in appropriate defcriptions, and figures, of the whole ENTO-MOLOGIA BRITANNICA, fo far as his own cabinet, and the obliging communications of his friends will permit. —And here the author begs leave to ftate, that the time elapfed fince the conclufion of the former part of the work has been employed in a manner beft calculated to give effect to this defign. Independently of many valuable acquifitions collected by himfelf and various friends in remote parts of the kingdom, he has the fatisfaction to obferve, that two entire cabinets, of eminent celebrity, have been lately added to that which he before poffeffed.

The

The firſt of thoſe collections is of the utmoſt conſe-
quence to the ſcientific Entomologiſt, as muſt be admitted, when it is obſerved to be the genuine cabinet
of Britiſh Inſects, formed by the late Mr. D. Drury, the
patron of Harris, and father, as he may be truly deemed,
of practical Entomology in this country :—a cabinet, the
reſult of thirty years, induſtriouſly and moſt ardently de-
voted to this purſuit, and combining the united information and diſcoveries of almoſt every other Engliſh Aurelian for a long period of time.—And even after naming
this, perhaps the firſt eſtabliſhed cabinet of any note in
England, it will not appear trivial to mention the other,
that of the late Mr. Green, of Weſtminſter, a collector
well known to the practical Entomologiſts of the preſent
day, as inferior to few, if any, in his zealous and perſe-
vering attachment to this ſubject. The poſthumous la-
bours of two or three other collectors might be likewiſe
named, as at this time enriching the author's cabinet, and
one eſpecially of Kentiſh Inſects, collected in the neighbourhood of Faverſham ; but enough the author preſumes
has been already ſaid, to prove that he has been no less
ſuccefsful than affiduous, in availing himſelf of ſuch pre-
eminent advantages, and that he has ultimately amaſſed together, ſuch a collection of the Entomological productions
of this country, as may enable him to render the continuation of his Natural History of Britiſh Inſects ſo far
reſpectable as to gratify every moderate ſhare of expectation.

And laſtly, the author truſts, that in thus proceed-
ing upon an enlarged and comprehenſive plan towards
the elucidation of this pleaſing department of Britiſh
Natural

Natural Hiftory, he will merit the liberal countenance, not only of every Entomologift, but of every friend to the purfuits of fcience; and be enabled, through their kind communications, to bring forward and complete, a more copious, interefting, and ufeful work, than can even now be anticipated.

THE

# 361

THE

# NATURAL HISTORY

OF

# BRITISH INSECTS.

═══════════

## PLATE CCCLXI.

### SPHINX CAROLINA.

YELLOW SPOTTED UNICORN HAWK-MOTH.

LEPIDOPTERA.

*GENERIC CHARACTER.*

Antennæ fomewhat prifm-formed, and thickeft in the middle: tongue moft commonly exferted: feelers two: wings deflected.

*SPECIFIC CHARACTER*

AND

*SYNONYMS.*

Wings clouded, entire, pofterior margin dotted with white: abdomen with five (or fix) pair of fulvous fpots.

SPHINX CAROLINA: alis integris omnibus margine poftico albo punctato, abdomnis ocellis fex parium fulvis. *Lin. Syft. Nat.* 2. 798. 7.—*Muf. Lud. Ulr.* 346.—*Gmel. Linn. Syft. Nat.* 2377. 7.

VOL. XI.                    B                    SPHINX

Sphinx Carolina. *Fabr. Fnt. Syft. T. 3. p.* 1 *p. 363. n.* 25.
Sphinx 5-maculatus, the yellow fpotted Unicorn. *Haw. Lep.*
   *Brit.* 59. *Sp.* 3.

———————

We are happy to embrace the prefent opportunity of prefenting our readers with a figure of this magnificent fpecies of Hawk-Moth, as a new Britifh Infect, upon the beft and moft unqueftionable authority. We have a fpecimen of it among the Britifh Sphinges, in the cabinet of the late Mr. Drury, now in our poffeffion, with a manufcript note affixed, informing us that this identical infect was taken in the neighbourhood of London, and brought to him alive fome few years ago *. The figure accompanying this defcription will afford a better idea of the beauty of this valuable acquifition, than any words we can employ; it is reprefented precifely in its natural fize, and as nearly refembling it in markings, and colours, as the fidelity of the pencil will admit.

When we fay the figure of this infect is fubmitted for the firft time as a Britifh fpecies, we wifh to be underftood as fpeaking of the figure only, for the very fpecimen under confideration at this time has been already defcribed as a Britifh infect, and the fpecies itfelf is perfectly well known as an exotic, or extra-european kind, to moft entomologifts. It is this fpecimen that Mr. Haworth mentions in his recent essay on the Lepidoptera of Great Britain, and upon the fole authority of which he inferted it in that work as a new Britifh

———————

* The label alluded to, refers to two fpecimens, namely, our prefent infect, and one of Sphinx Convolvuli, both which are mentioned in the following words, infcribed in the hand-writing of Mr. Drury. " One of the above fpecies is certainly different from the *Sph. Convolvulus.* The difference is manifeftly difcernable. They were alive when firft brought to me, one about the year 1776, the other 1788."—It is obviouily impoffible to collect from the tenor of this memorandum, which of the two infects he received firft, but this we may reft perfuaded of, that he obtained the living fpecimen of our new Britifh fpecies either in the year 1776, or 1788.

                                                          **Infect.**

Infect. We have, however, still further to observe, that although it was unique as *British* at the time Mr. Haworth described it from Mr. Drury's cabinet, it is not so at present, another collector, as Mr. Haworth informs us, having captured a specimen of it very lately in the vicinity of Little Chelsea, near which place it proves, upon pretty accurate information that Mr. Drury's specimen was also taken.

These are our authorities for considering the species as British, and of course as claiming a very distinguished place in the present work, not less on account of its magnitude, than its beauty and rarity. That it is occasionally found in Britain is sufficiently obvious, but there are circumstances attending its history that leave some doubts upon our mind, whether we ought not rather to consider it as a naturalized species, than as an *aborigine,* at the same time that the absolute impossibility of deciding this doubtful particular must be acknowledged.—In America, we well know, it is far from uncommon, and being naturally a hardy species, there is at least a possibility of the parent stock of the English brood having been originally introduced into this country with the cargoes of some American vessels.

This being the true Sphinx Carolina of Linnæus, an insect so very clearly ascertained both from the Linnean description of it, and from the figure quoted in the works of Merian, we cannot avoid expressing some surprise, that Mr. Haworth, in his recent publication above-mentioned, should have deemed it altogether a new species. The circumstance of Mr. Drury's specimen having only five pair of lateral spots on the abdomen, instead of six as Linnæus remarks in speaking of his Sphinx Carolina, may perhaps have led to this error; for in every other particular Linnæus is surely too expressive to be easily mistaken. So far as relates to the number of those yellow lateral spots, the Linnæan definition must be understood with some latitude, for Linnæus would certainly have been more correct in stating five spots on each side to be the usual number, instead of six. All the specimens of Sphinx Carolina that have occurred to our own observation, have been uniformly marked with five pair of la-

B 2                                                         teral

teral fpots only, with the exception of one or two large females, in which there was a flight appearance of a fixth pair ; a few fulvous hairs appearing below the black band on each fide the fixth annulation of the abdomen.

The larva of this infect is green, with lateral fpiracles on every fegment, furrounded by a purple ring, and the caudal fpine is of the fame colour.   According to Fabricius the larva feeds on the Tobacco plant: Mr. Abbot alfo confirms this fact in his hiftory of the Infects of New Georgia, fo that whatever it may fubfift upon in this country, we muft conclude the Tobacco plant to be its natural food.   In America we are informed, that it is really diftinguifhed by the name of *Tobacco Moth.*

PLATE

# 362

# PLATE CCCLXII

## FIG. I. I.

### COCCINELLA OBLONGO-GUTTATA.

OBLONG-SPOTTED LADY COW.

COLEOPTERA.

*GENERIC CHARACTER.*

Antennæ clavated, club folid: anterior feelers femicordated: thorax, and wing-cafes margined: body hemifphærical: abdomen beneath black.

*SPECIFIC CHARACTER*

AND

*SYNONYMS.*

Shells red; with lines and dots of white.

COCCINELLA OBLONGO-GUTTATA: coleoptris rubris: lineis punc-
tifque albis. *Linn. Syft. Nat.* 584. 38.—*Faun.
Suec.* 496.—*Gmel.* 1660. 38.
COCCINELLA OBLONGO-GUTTATA. *Fabr. Spec. Inf. I.* 103. 57.—
*Mant. I.* 60. 79.—*Ent. Syft. I. p.* 1. 296. 91.
*Maryh. Ent. Brit. T. 1. p.* 162. *fp.* 34.
*Degeer,* 5. 384. 19.
*Panz. Ent. Germ.* 146. 50.
*Schaeff. Icon. t.* 9. *f.* 10.

This

This appears on the credit of moſt writers to be a rare inſect. Our ſpecimen was taken in Kent. It is ſaid to inhabit the Pine, *Pinus ſylvestris.*

We ſhould in particular obſerve that the **prevailing or ground co**lour of the wing-caſes and thorax **in our** ſpecimen is not red as the ſpecies is uſually deſcribed, but rather of a light or teſtaceous brown, at the ſame time that its variegations of white marks and ſpots agree with the Linnæan deſcription of the inſect.

The ſmaller inſect at Figure I. is of the natural ſize.

---

# F I G. II. II.

## COCCINELLA TREDECIM-PUNCTATA

### 13-DOT LADY COW.

*SPECIFIC CHARACTER*

AND

*SYNONYMS.*

Shells yellow, or red, with thirteen black dots; body oblong.

Coccinella 13-Punctata : coleoptris luteis: punƈtis nigris tredecim, corpore oblongo. *Linn. Syſt. Nat.* 582. 20.—*Fn. Suec.* 481.—*Gmel. Linn. Syſt. Nat.* 1653. 20.—*Fabr. Syſt. Ent.* 83. 25.—*Spec. Inſ. I.* 99. 38.—*Mant. I.* 58. 54.—*Ent. Syſt. I. p.* 1. 279. 61.—*Marſh. Ent. Brit. T. I. p.* 156. *ſp.* 19.—*Panz. Ent. Germ.* 139. 27.—*Degeer. V.* 375. 9.

La coccinelle rouge à treize points noir, et corcelet rouge à bande. *Geoffr. Inſ. I. p.* 324. *ſp.* 7.

Linnæus,

# PLATE CCCLXII.

7

Linnæus, and after him Fabricius, and feveral other writers, defcribe this infect as having the fuperior furface yellow with black fpots. This is commonly the colour, but it alfo occurs pretty frequently of a reddifh as well as yellow colour, and even fometimes assumes a vermillion tint as brilliant as the common Lady Cow. *Coccinella feptem-punctata.* Geoffroy defcribes it as being of a red colour. This kind is found among plants ; is faid to inhabit *Armoracia.*

---

# FIG. III.

## COCCINELLA SEPTEM-NOTATA.

### SEVEN-DOT RED LADY COW.

*SPECIFIC CHARACTER*

AND

*SYNONYMS.*

Oblong : wing cafes red with feven black fpots on each : margin of the thorax and two dots white.

COCCINELLA 7-NOTATA: oblonga coleoptris rubris: punctis feptem nigris, thoracis margine punctifque duobus albis. *Fabr. Ent. Syft. I. p.* 1. 275. 43.
*Panz. Faun. Germ.* 187. 20.
*Marfh. Ent. Brit. T. I. p.* 153. *fp.* 11.
COCCINELLA MUTABILIS. *Payk. Faun. Suec.* 2. 39. 40.
COCCINELLA CONSTELLATA. *Laich.* 121. 6.

---

An elegant fpecies, and not very common. Its *habitat* unknown. This infect is evidently different from the Linnæan *Coccinella 7-punctata,* already figured in this work\*, but to which it bears a remote refemblance.

\* *Brit. Inf.* Pl. 39. f. 5.

FIG,

PLATE CCCLXII.

## FIG. IV. IV. V. V.

### COCCINELLA 24-PUNCTATA.

24-DOT RED LADY COW.

*SPECIFIC CHARACTER*

AND

*SYNONYMS.*

Wing-cafes red, with twenty-four black fpots.

COCCINELLA 24-PUNCTATA : coleoptris rubris: punctis nigris vi-
ginti quatuor. *Linn. Syft. Nat.* 583. 28.—*Fn.
Suec.* 487.—*Gmel. Linn. Syft. Nat.* 1655. 28.
COCCINELLA 24-PUNCTATA. *Fabr. Syft. Ent.* 84. 33.—*Spec.
Inf. I.* 101. 47.—*Mant. I.* 59. 66.—*Ent. Syft.
I. p.* 1. 281. 72.
*Marfh. Ent. Brit. T. I. p.* 159. *fp.* 26.
*Panz. Ent. Germ.* 142. 37.
La Coccinelle rayée, *Geoffr. Inf. I.* 326. *n.* 11.
*Degeer Inf. V.* 381. 14.

———————

Two diftinct varieties of this variable fpecies are figured in our
plate, fig 4 and 5, one of which has the black dorfal dots of a fmall
fize, the other large. This fpecies is commonly found on flowers.

PLATE

# 363

# PLATE CCCLXIII.

## FIG. I.

## PHALÆNA POTAMOGATA.

### CINEROUS CHINA MARK MOTH.

#### LEPIDOPTERA.

##### GENERIC CHARACTER.

Antennæ tapering from the bafe : wings in general deflected when at reft. Fly by night.

##### * GEOMETRA.

##### SPECIFIC CHARACTER

##### AND

##### SYNONYMS.

Wings cinereous, with white fpots : anterior pair obfoletely reti-culated.

PHALÆNA POTAMOGATA: feticornis alis cinereis albo maculatis : anticis obfolete reticulatis. *Linn. Syft. Nat.* 2. 873. 275.—*Fn. Suec.* 1299.
*Fabr. Ent. Syft. T. 3. p. 2. p.* 213. *fp.* 313.

The larva of this fpecies is fuppofed to feed principally on the *Pota-mogaton natans* from which circumftance it has been called fpeci-cally Potamogata. It appears early in the month of June in the

VOL. XI.                         C                         winged

winged state hovering about aquatic plants in ditches, and other watery places. This is a very common species, and is frequently found drowned, and lying on the surface of the water where aquatic plants are abundant.

## FIG. II.

## PHALÆNA STAGNATA.

PEARL CHINA MARK.

*SPECIFIC CHARACTER.*

PHALÆNA STAGNATA : wings white, with two irregular common subfuscous bands; the outer one furcating from the middle of the anterior wings to the costal margin.

The general colour of this insect is a beautiful delicate white, with a perlaceous nue. The transverse fuscous bands are so disposed on the anterior wings as to give it somewhat of a reticulated appearance, but less so than in Phalæna Potamogata, and several other species of *China-marks*, as English collectors denominate them. The bands on the posterior wings are not in any manner reticulated. This does not appear to be a very common species.

PLATE

# 364

# PLATE CCCLXIV.

## FIG. I. I.

## PHALÆNA ARCUANA.

### CURVE-BANDED TORTRIX-MOTH.

#### LEPIDOPTERA.

*GENERIC CHARACTER.*

Antennæ gradually tapering from the bafe to the tip: wings in general deflected when at reft. Fly by night.

\* TORTRIX.

*SPECIFIC CHARACTER*

AND

*SYNONYMS.*

Wings yellowifh-brown, with three filvery curved bands; and a black fpot in the difk, on which are three filvery dots.

TORTRIX ARCUANA: alis luteis: fafciis tribus arcuatis maculaque difci atra; punctis tribus argenteis. *Linn. Syft. Nat. 2.* 877. 296. *Fn. Sv.* 1317.

PHALENA ARCUANA. *Fabr. Sp. Inf. 2. p. 281. n. 31. Mant. Inf. 2. p. 230. n. 53.—Ent. Syft. T. 3, p. 2. p. 260. n. 72.*
     *Clerk-Phal. tab. 10. fig. 2.*

C 2                                            Phalena

Phalena arcuana is an infect of uncommon beauty. The general colour of the anterior wings is yellowifh brown, or teftaceous, varied with darker towards the exterior margin, and tranfverfely ftriped with filvery: there are alfo at the bafe two remarkable arched, or incurvated filvery lines. In the difk, a little inclining towards the inner margin, is a broad fpace, of a pale yellow colour, in the center of which is a black fpot, enriched with three filvery dots. The lower wings are obfcure.

This infect is found on the nut tree in its perfect ftate: its tranf-formations are not clearly known.

---

## F I G. II. II.

## PHALÆNA DIMIDIANA.

### BROWN AND ORANGE WING TORTRIX-MOTH.

#### SPECIFIC CHARACTER.

TORTRIX DIMIDIANA: anterior half of the firft wings fufcous, pofterior teftaceous-orange, with four fhort filvery lines on the exterior margin.

---

This little moth, which we are inclined to confider as an undefcribed fpecies, is little more than one third the fize of the preceding infect. The fufcous, and rich teftaceous-orange of the anterior wings, appear perfectly diftinct and independent of each other at the bafe and apex of the wing, but unite and blend together about the middle, or a little inclining towards the pofterior end: the whole furface has a flightly gilded, or metallic glofs.

FIG.

## FIG. III. III.

## PHALÆNA NEBULANA.

BLACK-CLOUDED TORTRIX-MOTH.

*SPECIFIC CHARACTER.*

Tortrix nebulana: anterior wings fub-teftaceous, and varied: the clouds in the difk, and marginal fpots, deep fufcous.

———————————

Taken in Darent Wood, Dartford, in July.   This is a new fpecies.

PLATE

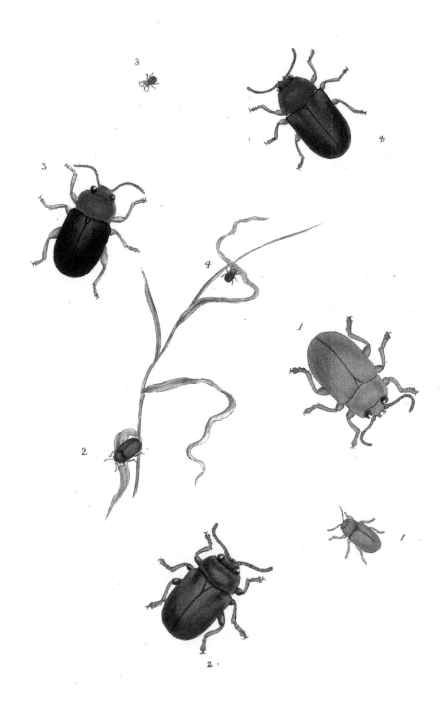

# 365

# PLATE CCCLXV.

## FIG. I. I. II. II.

## CHRYSOMELA GRAMINIS.

GREAT GRAMINIFEROUS CHRYSOMELA.

COLEOPTERA.

*GENERIC CHARACTER.*

Antennæ moniliform, thicker at the extremity: head inferted: thorax marginate: wing-cafes immarginate: body in general ovate and convex.

*SPECIFIC CHARACTER*

*AND*

*SYNONYMS.*

Green-blue, polifhed: antennæ and legs fame colour.

CHRYSOMELA GRAMINIS: viridi-cærulea nitida, antennis pedibufque concoloribus. *Linn. Fn. Suec.* 509.—*Linn. Syft. Nat.* 587. 7.—*Gmel. Linn. Syft. Nat.* 1670. 7.

CHRYSOMELA GRAMINIS. *Fabr. Syft. Ent.* 96. 9.—*Sp. Inf. I.* 118. 16.—*Mant. I.* 68. 21.—*Ent. Syft. I. p.* 2. 314. 33.

CHRYSOMELA GRAMINIS. *Marfh. Ent. Brit. T. I. p.* 172. *fp.* 6.
Le grand Vertubleu *Geoffr. I.* 260. 10.

Two

Two very remote kinds, or varieties, of Chryfomela graminis, are fhewn in the prefent plate ; the firft is of the ufual fort, green gloffed with gold, and which in fome fpecimens is of inimitable fplendour : the other is fmaller, and of a rich and deep blue, fimilar to fome individuals of Chryfomela Alni, but from which it may be at once diftinguifhed by being deftitute of the impreffed, or excavated dots, on the wing-cafes fo apparently in that fpecies ; and by having the legs and antennæ partaking of the fame colour as the reft of the body, while thofe parts in Alni are black.

According to Geoffroy, Chryfomela Graminis is found on the Galeopfis, Lamium, and other labiofe plants. Linnæus names it fpecifically graminis from its feeding upon grafs, in which particular he has been followed by moft writers. The fmaller figures 1 and 2, denote the natural fize.

---

## F I G. III. III.

## CHRYSOMELA RUFICORNIS.

### RUFOUS-HORNED CHRYSOMELA.

\* *Section* Altica *posterior Thigh very thick,*

*SPECIFIC CHARACTER*

AND

*SYNONYMS.*

Blue ; head, thorax, antennæ, and legs rufous : wing-cafes with crenate ftriæ.

CHRYSOMELA RUFICORNIS: cærulea, capite thorace antennis pedibufque rufis, elytris crenato-ftriatis. *Marfh. Ent Brit. T,* I. *p.* 199. 70.

GALLERUCA

# PLATE CCCLXV. 17

GALLERUCA RUFICORNIS: cærulea capite thorace antennis pedibufque rufis, elytris crenato-ftriatis. *Fabr. Ent. Syft. I, p. 2. 32. 96.*
*Panz. Faun. Germ. 21. 12.*

ALTICA RUFICORNIS *Panz. Ent. Germ. 179. 19.*

CHRYSOMELA cæruleo-ftriata *De Geer V. 343. 48.*

Habitat of this little fpecies unknown.

# FIG IV. IV.

## CHRYSOMELA RUFIPES.

### RUFOUS-LEGGED CHRYSOMELA.

*SPECIFIC CHARACTER*

AND

*SYNONYMS.*

Oblong : blue ; head, thorax, legs, and antennæ rufous.

CHRYSOMELA RUFIPES : cærulea obovata, capite thorace pedibus antennifque rufis. *Linn. Syft. Nat. 595. 65,—Faun. Suec. 545,—Gmel. Linn. Syft. Nat. 1695. 65.*

ALTICA RUFIPES. *Fabr. Syft. Ent. 114. 14.*

GALLERUCA RUFIPES. *Fabr. Ent. Syft. I. p. 2. 32. 94.*

CHRYSOMELA RUFIPES. *Marſh. Ent. Brit. T. I. p.* 198. *ſp.* 68.
*De Geer Inſ.* 5. 343. 47. *t.* 10. *f.* 11.
*Panz. Ent. Germ.* 179. 17—*Faun. Germ.* 21. *t.* 10.

This is a ſmall and rather uncommon ſpecies.   Taken in Kent.   Inhabits plants,

PLATE

366

PLATE CCCLXVI. 19

# PLATE CCCLXVI.

## MUSCA PULCHELLA.

### STRIPED-WING MUSCA.

#### DIPTERA.

*GENERIC CHARACTER.*

Mouth with a foft exferted flefhy probofcis, and two unequal lips: fucker befet with briftles: feelers fhort, and two in number, or fometimes none: antennæ ufually fhort.

\* *Section, Antennæ a naked briftle.*

*SPECIFIC CHARACTER*

AND

*SYNONYMS.*

Downy, cinereous: difk of the wings yellowifh-brown, with a flexuous white hyaline ftripe.

MUSCA PULCHELLA: antennis fetariis pilofa cinerea alarum difco fufco flavefcente: vitta flexuofa albo-hyalina. *Fabr. Ent. Syft. T. 4. p. 352. fp.* 167.

Mufca pulchella antennis fetariis pallida teftacea pilofa alis patulis late flavo nigro fafciatis. *Roffi. Fn. Etrufc.* 2. 314. 1528. *tab.* 8. *fig.* 6. *mal.*

The Fabrician Entomological work above-mentioned, affords a copious and diftinct account of this elegant fpecies of Mufca. Fabricius met with it in the cabinet of M. Bofc, and obferves that it inhabits Gardens in Italy. Two years previous, however, to the appearance of *Entomologia Syftematica\**, Roffius had defcribed and figured this fpecies in his *Fauna Etrufca* †, as an Italian infect, fo that the latter muft be confidered as the firft defcriber of it. Probably it has not been noticed by any other continental writer fince ‡ : as a native of Great Britain it is certainly undefcribed.

Mufca pulchella we muft efteem as a very fcarce infect in this country. Our fpecimens were taken in the Wilds of Kent, near Faverfham, and it has occurred, though rarely, as we are informed, nearer the vicinity of London.

The upper figure in the plate exhibits an enlarged reprefentation of this curious infect in a flying pofition, the natural fize appears below.

---

\* 1792.

† Publifhed in 2790.

‡ Gmelin omits this and many other very interefting infects defcribed by Fabricius, which we might expect to find in his improved edition of the Linnæan Syftema Naturæ.

PLATE

PLATE CCCLXVII.                    21

# 367

# PLATE CCCLXVII.

## FIG. I. I.

## CARABUS PILICORNIS.

### HAIRY-HORNED CARABUS.

#### COLEOPTERA.

*GENERIC CHARACTER.*

Antennæ filiform: feelers fix, the exterior joint obtufe and truncated: thorax obcordated, truncated behind, and margined: wing-cafes margined: abdomen ovate.

*SPECIFIC CHARACTER*
AND
*SYNONYMS*

Thorax roundifh: wing-cafes ftriated, with impreffed dots: antennæ hairy.

CARABUS PILICORNIS: thorace rotundato elytris ftriatis punctifque impreffis, antennis pilofis. *Fabr. fp. Inf.* 1. *p.* 307. *n.* 48.—*Mant.* 1. 200. 65.—*Ent. Syft.* 1. *p.* 1. 152. 122.

CARABUS PILICORNIS. *Marfh. Ent. Brit. T.* 1. *p.* 446. *fp.* 36. *Panz. Faun. Germ.* 11. *t.* 10.

CARABUS PILICORNIS. *Donov. Tour South Wales, V.* 1. *p.* 380.

This infect appears to be rare in England. The firft fpecimen of it met with by ourfelves was taken on the fandy fhore of the Severn fea,

fea, near the village of Newton, Glamorganfhire: another occurs in the cabinet of the late Mr. Green, now in our poffeffion, but the habitat of the latter is unknown to us.

---

## FIG. II. II.

## CARABUS SEMIPUNCTATUS.

### HALF-DOTTED CARABUS.

#### SPECIFIC CHARACTER.

CARABUS SEMIPUNCTATUS: thorax roundifh: wing-cafes fuscous, ftriated, with anterior hyaline fpots, and dots of the fame on the pofterior half.

CARABUS SEMIPUNCTATUS. *Donov. Tour South Wales, V. I. p.* 380.

---

We found a fpecimen of this curious fpecies in the fame place, and at the fame time as the preceding. It is not defcribed by any author.

PLATE

PLATE CCCLXVIII. 23

# 368

# PLATE CCCLXVIII.

## SYNODENDRON CYLINDRICUS.

### CYLINDRICAL SYNODENDRON.

#### COLEOPTERA.

*GENERIC CHARACTER.*

Antennæ lamellated: palpi four, equal: lip filiform, horny, palpigerous at the tip: body cylindrical, obtufe at both extremities: anterior fhanks dentated.

*SPECIFIC CHARACTER*

AND

*SYNONYMS.*

Anterior part of the thorax truncated, and five-toothed: an erect horn on the head.

SYNODENDRON CYLINDRICUM: thorace antice truncato quinque dentato, capitis cornu erecto. *Fabr. Ent. Syft. T. 1. p. 2. 358. 94. n. 1. Paykull Faun. Suec. 111. 140. 1. Panz. Ent. Germ. 282. 1. Fueft. Archiv. 67. 4.*

SCARABÆUS CYLINDRICUS. *Linn. Syft. Nat. 544. 11.—Faun. Suec. 380.—Gmel. 1532. 11.*

LUCANUS CYLINDRICUS. *Laich. Inf. Tyr. 3. 4. Marfh. Ent. Brit. T. 1. 50. 4.*

Lucanus Tenebroides. *Scop. Ann. 5.—Nat. Hift. 10.*

The male of this fpecies is fufficiently diftinguifhed by the erect horn on the anterior part of the head, the female being deftitute of this character: in other refpects they nearly correfpond. Both fexes are reprefented in their natural fize on the oppofite plate. Lives in the trunks of trees, Inhabits various parts of Europe.

PLATE

# 369

# PLATE CCCLXIX.

## FIG. I. I.

### PHALÆNA TRIMACULANA.

THREE-SPOT TORTRIX-MOTH.

LEPIDOPTERA.

*GENERIC CHARACTER.*

Antennæ gradually tapering from the bafe to the tip : wings in ge-neral deflected when at reft. Fly by night.

* TORTRIX.

*SPECIFIC CHARACTER.*

TORTRIX TRIMACULANA : anterior wings teftaceous and fufcous varied : a pale angular tranfverfe band near the bafe ; and whitifh fpace, inclofing three fmall dark dots behind.

———

A pretty fpecies, and moderately large. The ground colour chiefly teftaceous, varying from pale to darker in different fpecimens. The broad tranfverfe angular band, and fpot inclofing three fmall dots behind, are fufficiently characteriftic of this infect. It has alfo feveral flender whitifh lines, difpofed obliquely at the outer edge of the wing, and at the apex a fmall fubocellated fpot. We have not obferved any defcription either of this, or the third fpecies reprefented in our plate 369, in any work, the fecond fpecies appears in Hubner's *Beiträge zur Gefchichte der Schmetterlinge, &c.*

VOL. XI. E FIG.

## FIG. II. II.

## PHALÆNA BETULANA.

ALDER TORTRIX-MOTH.

*SPECIFIC CHARACTER.*

TORTRIX BETULANA : anterior wings ochraceous, with an oblique,
fubfufcous band acrofs the middle, and two black
dots : one central and touching the band.

PHALÆNA BETULANA. *Hubn. Beitr.*

This is one of the larger fpecies of the Tortrix tribe; the colour ochraceous, fometimes livid, at others tinged with reddifh, and
gloffy. It may be readily diftinguifhed by the oblique dark band
acrofs the middle of the wing, to which one of the black fpots is
connected ; the other dot is fmaller, and placed nearer the pofterior end
of the wing. There is alfo a ferruginous dafh contiguous, that extends to the outer margin of the wing.

FIG

# PLATE CCCLXIX.

27

## FIG. III. III.

### PHALÆNA NOTANA.

DOTTED TORTRIX-MOTH.

*SPECIFIC CHARACTER.*

TORTRIX NOTANA: anterior wings fubferruginous, with numerous diftinct black dots.

———————

Taken in Darent Wood, Dartford, in July, and alfo in Coombe
Wood, Surrey.

E 2                    PLATE

# 370

# PLATE CCCLXX.

## FIG. I. I.

## PHALÆNA GEMINANA.

### DOUBLE-DOT TORTRIX MOTH.

#### GENERIC CHARACTER.

Antennæ gradually tapering from the bafe to the tip : wings in general deflected when at reft.  Fly by night.

#### * TORTRIX.

#### SPECIFIC CHARACTER.

TORTRIX GEMINANA.  Anterior wings pale with a broad fufcous ftripe along the middle, edged interiorly with a jagged whitifh line : a fmall teftaceous fpot, with two black dots near the anal angle.

———

The prevailing colour of the upper wings in this fpecies, when the infect is in perfect condition, is of milky yellowifh, varied with teftaceous.  Befides the broad fufcous longitudinal fhade, and teftaceous double dotted fpot behind, as mentioned in the fpecific character, there are a variety of elegant markings and lineations of teftaceous brown and black at the apex and along the outer edge.  We fufpect that it is an uncommon infect, having hitherto only met with it once : —this was taken in Kent.

FIG.

# PLATE CCCLXX.

## FIG. II. II.

## PHALÆNA TRIFASCIANA.

THREE BANDED TORTRIX MOTH.

*SPECIFIC CHARACTER.*

Wings whitifh, with three brown bands margined with black dots,

TORTRIX TRIFASCIANA: alis albis: fasciis tribus fufcis; tertia nigro punctata. *Fabr. Ent. Syft. T. 3. p. 2. p. 248. fp. 25.*

There can fcarcely remain the flighteft doubt of this being the fpecies of Tortrix defcribed by Fabricius from the cabinet of **Dr.** Allioni, under the name of trifafciana, prefuming however that the infect Fabricius faw, muft have been in lefs perfect condition than our fpecimen: he defcribes the fituation of the three bands very exactly, one at the bafe of the wing, the fecond oblique acrofs the middle, and the third at the tip, the laft of which he obferves are dotted with black. To this we may add, that when perfect, all the bands are circumfcribed within a double feries of black dots, although thofe on the brown ftripe at the tip are commonly moft confpicuous. Taken in Coombe Wood, Surrey.

**PLATE**

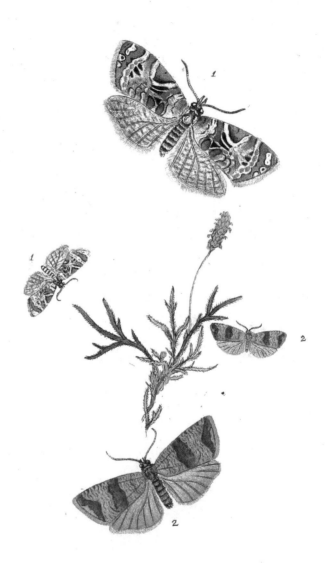

# PLATE CCCLXXI.

## FIG. I. I.

### PHALÆNA V-ALBANA.

WHITE-V TUFTED TORTRIX-MOTH.

LEPIDOPTERA.

*GENERIC CHARACTER.*

Antennæ gradually tapering from the bafe to the tip : wings in ge-neral deflected when at reft.   Fly by night

\* TORTRIX.

*SPECIFIC CHARACTER.*

Anterior wings brownifh, variegated with pale rivofe lines and tufted dots, and a white flexuous V-like mark at the coftal margin.

TORTRIX V-ALBANA. *Marfh. M.S. Ent. Brit.*

---

This is a charming little fpecies, and very far from common.   The ground colour of the anterior wings is pale fufcous, and the rivofe lines that variegate it whitifh, with a tinge of teftaceous brown, or reddifh difpofed chiefly in dots along the middle.   But the moft confpicuous mark, and by means of which this fpecies of tufted tortrix may be eafily known, is the white coftal flexuous band in this middle of the anterior wing which bears a ftrong refemblance to the letter V.   The pofterior wings are pale with numerous fhort dafhes, or interrupted tranfverfe darker lines.

FIG.

# FIG. II. II.

## PHALÆNA BILITURANA.

### DOUBLE-BANDED TORTRIX-MOTH.

*SPECIFIC CHARACTER.*

Anterior wings cinereous brown, with a fufcous band acrofs the middle, and another fubterminal at the pofterior end.

TORTRIX FASCIANA. *Fabr. Ent. Syſt. T.* 3. *p.* 2. *p.* 261. *n.* 782?

———

This infect approaches very nearly to Phalæna fafciana of Fabricius: it is alfo allied to Phalæna Gerningana of the fame author, and it is not unlikely, on future inveftigation, they may both prove to be accidental varieties of the fame fpecies as our infect. The upper wings in our fpecimen is of a cinereous brown colour inclining to reddifh, and marked with many fhort tranfverfe lines. Acrofs the middle is a broad band, and at the tip another fmaller one, with a flexuous edge, leaving a pale narrow fpot in the middle of the pofterior apex next the margin.

Taken in Kent near Faverfham.

PLATE

372

# PLATE CCCLXXII.

## MUSCA PLUVIALIS.

### RAINY FIVE-SPOT MUSCA.

#### DIPTERA.

##### GENERIC CHARACTER.

Mouth with a foft, exferted, flefhy probofcis, and two unequal lips: fucker befet with briftles: feelers fhort, and two in number, or fometimes none: antennæ ufually fhort.

\* *Antennæ a naked briftle.*

##### SPECIFIC CHARACTER

###### AND

###### SYNONYMS.

Cinereous, with five black fpots on the thorax, and obfolete fpots on the abdomen.

MUSCA PLUVIALIS: cinerea, thorace maculis quinque nigris, abdomine obfoletis. *Linn. Fn. Suec.* 1844.—*Linn. Syft. Nat.* 2. 992. 83.—*Gmel. Linn. Syft. Nat. T.* 1. *p.* 5. 2847. *fp.* 83.

MUSCA PLUVIALIS. *Fabr. Spec. Inf.* 2. *p.* 443. *n.* 40.—*Mant. inf.* 2. *p.* 346. *n.* 47.—*Ent. Syft. T.* 4. *p.* 329. *Sp.* 71.

La Mouche cendrée à points noirs. *Geoffr. Inf. Par.* 2. *p.* 529. *n.* 68. *De Geer. Inf.* 6. *p.* 27. *n.* 6.

VOL. XI.        F        This

This pretty infect is a general inhabitant of Europe. Before rain it is obferved to affemble in fwarms, and conceal itfelf under the leaves of plants, where it remains perfectly tranquil till the rain is over. It is reprefented both in the natural fize, and magnified, in the annexed plate.

PLATE

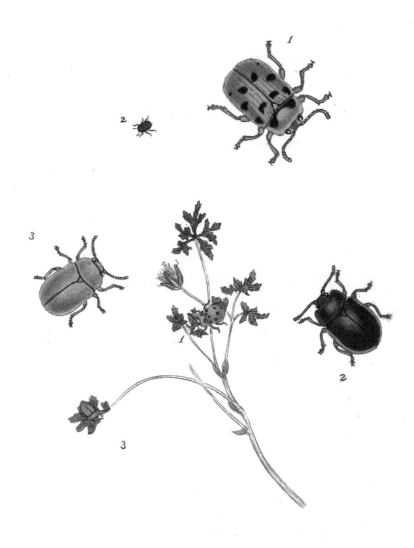

# 373

# PLATE CCCLXXIII.

## FIG. I. I.

## CHRYSOMELA 10-NOTATA.

### YELLOWISH TEN-DOT CHRYSOMELA.

### COLEOPTERA.

*GENERIC CHARACTER.*

Antennæ moniliform, thicker at the extremity: head inferted: thorax marginate, wing-cafes immarginate: body in general ovate and convex.

*SPECIFIC CHARACTER*

AND

*SYNONYMS.*

Yellow or yellowifh: thorax with two fub-conneĉted black dots, and five on the wing-cafes: legs yellowifh, or fubrufous.

CHRYSOMELA 10-NOTATA: flava, thoracis punĉtis fub-connexis duobus elytrorumque quinque nigris. *Marjh. Ent. Brit. T.* 1. *p.* 175. *fp.* 13.

Chryfomela 10-punĉtata β var. *Linn. Syft. Nat.* 590. 32.

Chryfomela rufipes. *De Geer V.* 295. 4. *t.* 8. *f.* 25.

La Chryfomele rouge à points noirs. *Geoffr. Inf. I.* 258. 4.

F 2                                     This

This infect is feparated from the Linnæan Chryfomela 10-punctata upon the authority of *Entomologia Britannica* as above quoted There is much reafon to believe it a diftinct fpecies, though we cannot fpeak precifely to that effect, fince it is poffible it may prove, on future obfervation, to be the female of Chryfomela 10-punctata, or a variety of it.   The two infects refemble each other in fize, and moft other particulars, the bilobate black mark, or confluent fpots on the thorax, and the colour of the mouth, and legs excepted: thofe of C. 10-punctata being black, while in our infect, they are conftantly yellow, or yellowifh-red, inclining to rufous.   Several writers agree that Chryfomela 10-punctata, is liable to much variation: Fabricius, in particular, tells us, he has obferved it with both the wing-cafes deftitute of the fifth, or pofterior fpot.   The lower furface is black.   Found on the afpin and willow.

The fmaller figures, as ufual, point out the natural fize of the infects reprefented in this plate.

---

## F I G. II. II.

## CHRYSOMELA AUCTA.

RED-BORDERED BLUE-CHRYSOMELA.

*SPECIFIC CHARACTER*

AND

*SYNONYMS.*

Blue, thorax polifhed: wing-cafes dotted, with a red margin.

CHRYSOMELA AUCTA: cyanea, thorace nitido, elytris punctatis: margine rubro. *Marfh. Ent. Brit. T. I. p.* 181. *fp.* 24

CHRYSOMELA

PLATE CCCLXXIII. 37

Chrysomela aucta: ovata thorace cyaneo nitido, elytris punctatis cyaneis: margine rubro. *Fabr. Mant. I.* 72.
69. *Ent. Syft. I. p.* 1. *326. fp.* 94.

Chryfomela aucta. *Gmel. Linn. Syft. Nat.* 1680. *fp.* 128.

Chrysomela marginata. *Act. Nidrof.* 3. 390. 80.

Firft defcribed by Fabricius from the cabinet of Zfchuck. The general colour above is a very deep purplifh blue, inclining almoft to
black, the margin of the wing-cafes excepted, that part being red:
the lower furface, together with the legs, are black.

## FIG. III. III.

## CHRYSOMELA HYPOCHÆRIDIS.

**CAT'S-EAR** CHRYSOMELA.

*SPECIFIC CHARACTER*

AND

*SYNONYMS.*

Entirely golden-green and polifhed.

Chrysomela Hypochæridis: tota viridi-aurata nitida. *Marfh.*
*Ent. Brit. T. I. p.* 184. *fp.* 35.

Chrysomela Hypochæridis: aurata, antennis nigris, elytris ab-
breviatis. *Linn. It. fcan.* 210.—*Faun. Suec.* 516.
—*Linn. Syft. Nat.* 589. 21.—*Gmel. Linn. Syft.*
*Nat.* 1675. 21.

I veftis Syngenefia. *Scop. Ent. Carn.* 193.

Linnæus

Linnæus confiders the colour of the antennæ in his fpecifical diftinction of this fpecies; thefe, he fays, are black, but it appears they are
not uniformly fo, being fometimes green. The fame infect has occafionally occured, likewife, of a green colour, without a golden glofs.
When fine, the golden coloured variety is a beautiful infect. Found
on the flowers of *Hypochæris maculata.*

PLA.

374

# PLATE CCCLXXIV.

## FIG. I.

### PHALÆNA LUNDANA.

ARCUATED TORTRIX-MOTH.

LEPIDOPTERA.

*GENERIC CHARACTER.*

Antennæ gradually tapering from the bafe to the tip: wings in general deflected when at reft.   Fly by night.

\* TORTRIX.

*SPECIFIC CHARACTER*

AND

*SYNONYMS.*

Wings at the bafe fufcous, with a pale femicircular ftripe: tip gloffed with gold, and ftreaked at the thicker margin with filvery, and yellow.

PHALÆNA LUNDANA: alis bafi fufcis: linea femicirculari pallida, apice auratis: margine craffiori argenteo flavoque ftrigato. *Fabr. Spec. Inf.* 2. *p.* 287. *n.* 74.—*Ent. Syft. T.* 3. *p.* 2. 282. *fp.* 166.

PHALÆNA BADIANA. *Wien. Verz.* 136. 8.

━━━━━━

This is an elegant, though fmall fpecies, which we have found during the fummer not uncommonly in the woods near the vicinity of London.   The fmallest figure i. denotes the natural fize.

# F I G.  II.

## PHALÆNA OBSCURANA.

FERRUGINOUS CLOUDED TORTRIX MOTH.

*SPECIFIC CHARACTER.*

Tortrix Obscurana.   Anterior wings fomewhat ferruginous, obfcurely clouded and fpeckled with fufcous: pofterior wings pale.

———————————

This appears to be an undefcribed fpecies: it is reprefented both in its natural fize, and magnified, in the oppofite plate.

PLATE

375

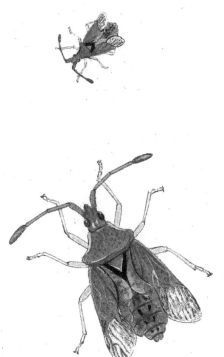

# PLATE CCCLXXV.

## CIMEX VENATOR.

### FAWN-COLOURED CLOVER BUG.

#### HEMIPTERA.

##### GENERIC CHARACTER.

Snout inflected: antennæ longer than the thorax: wings four, folded crofs-wife, anterior part of the upper pair coriaceous: back flat: thorax margined: legs formed for running.

*Section* Coreus. *Thorax spinous, body oblong, flat and deprefled: antennæ of four articulations, the exterior joint diftinctly ovate.*

##### SPECIFIC CHARACTER

###### AND

##### SYNONYMS.

Thorax obtufely fpined, obfcure grey: beneath yellowifh: antennæ and legs ferruginous.

Cimex venator: thorace obtufe fpinofo obfcure grifeus fubtus flavefcens antennis pedibufque ferrugineis. *Fabr. Ent. Syft. T.* 4. *p.* 128. 4.

———

We once met with the two fexes of this uncommon fpecies of Cimex crawling on a bed of clover, in a fmall field on one fide of Darent Wood, near Dartford, in Kent. It has occurred likewife to our obfervation in Surrey, and in the maritime parts of South Wales,

VOL. XI.    G    but

but that fo rarely that we are led to confider it as a fcarce infect, or at leaft as a very local one.

The only writer who has defcribed this Cimex, to our knowledge, is Fabricius, who faw an Italian fpecimen of it in the cabinet of Dr. Allioni, and introduced it to the notice of the Entomologift in his *Entomologia Syftematica*, under the title of Coreus Venator. No figure has hitherto appeared of this infect; nor has it been before mentioned as a native of any other part of Europe than Italy.

Both the upper and lower furface of Cimex Venator is fhewn in their natural fize, and an enlarged figure of the former in the center of the plate.

**PLATE**

# 376

# PLATE CCCLXXVI.

## FIG. I. I.

### VESPA DECIM-MACULATA.

TEN-SPOTTED WASP.

HYMENOPTERA.

*GENERIC CHARACTER.*

Mouth horny, with a compreffed jaw: feelers four, unequal, and filiform: antennæ filiform, the firft joint longeft and cylindrical; eyes lunar: body glabrous: fting pungent, and concealed within the abdomen. Both fexes have the upper wings folded.

*SPECIFIC CHARACTER.*

VESPA DECIM-MACULATA. Black: thorax immaculate: fcutel bidentated: firft five fegments of the abdomen, with a fubmarginal white dot on each fide.

---

Nearly allied to the Linnæan *Vefpa uniglumis* in point of fize, and general appearance, but differing in one very effential particular: the number of white fpots on the abdomen. The Vefpa uniglumis, (Crabro uniglumis of Fabricius) has white marginal dots only on three fegments of the abdomen; while, in our fpecies, the firft five fegments have a very confpicuous white dot on each fide. There are three or four other fpecies of Linnæan Vefpa, defcribed by Fabricius, in his new genera Crabro and Philanthus, that feem to bear fome refemblance to this infect, but which, on comparifon, appear to be certainly diftinct.

G 2                                                    The

The head and thorax are black: body of the fame colour, gloffy, and fpotted with white: legs yellow: thighs black. The only fpecimen we have yet met with of this kind, was taken in Kent. The fmalleft figure denotes the natural fize.

---

## FIG. II.

## APIS SPHECOIDES.

### SPHEX-FORMED BEE.

### GENERIC CHARACTER.

Mouth horny: jaw and lip membranaceous at the tip: tongue inflected: feelers four, unequal, filiform: antennæ fhort, and filiform, thofe of the female fomewhat clavated: fting of the females and neuters pungent, and concealed within the abdomen.

### SPECIFIC CHARACTER

### AND

### SYNONYMS.

APIS SPHECOIDES.   Deep black: abdomen ferruginous; bafe, and
tip black : wings blackifh.

MELITTA SPHECOIDES.   Aterrima; abdomine ferrugineo, bafi
apiceque nigro; alis nigricantibus.   *Kirby Ap.
Angl. T. 2. p. 46. fp. 9.*

SPHEX GIBBA : nigra; abdomine ferrugineo apice fufco: alis primoribus apice nigricantibus.   *Linn. Fn. Suec. 1658.—
Gmel. Linn. Syft. Nat. T. I. p. 5. p. 2732.*

SPHEX GIBBA : nigra abdomine ferrugineo apice fufco, alis anticis
apice fufcis.   *Linn.—Fabr. Ent. Syft. T. 2. p. 212.
n. 59.*

It

# PLATE CCCLXXVI. 45

It will be obferved, that the fpecific defcription of the Linnæan *Sphex gibba,* left us by Linnæus himfelf, does not very clearly exprefs our infect, but which it now appears, upon the beft authority, is certainly the one intended by that writer. This obfcure circumftance has been cleared up by Mr. Kirby, who met with the remains of the authentic fpecimen of Sphex gibba in the Linnæan cabinet, and was, by that means, enabled to afcertain the fpecies meant by Linnæus, which otherwife might have ever remained a matter of uncertainty. Fabricius, unacquainted with the infect, or more probably unable to determine the Linnæan infect from the defcription, is content to quote the words of Linnæus. Mr. Kirby has affigned it a new character, by which the fpecies may, in future, be eafily diftinguifhed. In his arrangement, it ftands as a *Melitta,* under the fpecific name of Sphecoides.

This infect is rather rare, our fpecimen was taken in Kent.

---

## FIG. III.

## APIS GEOFFRELLA.

### GEOFFROY'S BEE.

#### *SPECIFIC CHARACTER.*

Black: abdomen rufous, tip and legs black: tarfi and anterior fhanks reddifh.

APIS GEOFFRELLA. Atra; abdomine rufo, apice pedibufque nigro-piceis; maxillis, tarfis, tibiifque anticis, rufefcentibus. *Kirby. Ap. Angl. T.* 2. *p.* 45.

Geoffroy,

Geoffroy, in his *Hift. des Infeles*, mentions this infect as a fuppofed variety of his *Abeille noire à ventre brun et liffe*.   On the contrary, however, we are rather inclined to agree with Mr. Kirby, in believing it to be a diftinct fpecies.   Its fize is nearly the fame as the foregoing.

---

## F I G.  IV.

## APIS PUNCTATA

DOTTED BEE.

*SPECIFIC CHARACTER.*

Black with cinereous down : abdomen black, fegments with a white dot at each fide.

Apis punctata : nigra cinereo villofa abdomine atro : fegmentis
utrinque puncto albo.   *Fabr. Ent. Syft. T.* 2. 336.
*n.* 99.

Apis punctata.   *Kirby Ap. Angl. T.* 2. *p.* 219. *fp.* 35.

---

Defcribed by Fabricius as a native of this country.   The figure re-prefents it in the natural fize.

FIG.

# FIG. V.

## APIS FLORALIS.

FLORAL BEE.

*SPECIFIC CHARACTER.*

Entirely yellowifh-rufous: abdomen fpotted and fafciated with black.

APIS FLORALIS : tota rufa, abdominis fafciis quatuor vel quinque nigris. *Scop. Ann. Hift. Nat.* 4. *p.* 12. *n.* 7.——— *Gmel. Linn. Syft. Nat.* 2785. 125.

APIS FLORALIS : hirfuto flavefcens ; thorace fulvo ; abdomine maculis, fafciifque atris. *Kirby Ap. Angl. T.* 2. *p.* 324. *n.* 76.

————————

This fpecies of Bee is common, chiefly frequenting flowers.

**PLATE**

# 377

# PLATE CCCLXXVII.

## PHALÆNA RHEDIELLA.

### RHEDI'S TINEA-MOTH.

*GENERIC CHARACTER.*

Antennæ gradually tapering from the bafe to the tip : wings in ge-neral deflected when at reft. Fly by night.

\* TINEA.

*SPECIFIC CHARACTER*

AND

*SYNONYMS.*

Wings black : tip fulvous, with interrupted filver streaks.

PHALÆNA RHEDIELLA : alis nigris apice fulvis : ftrigis interruptis argenteis. *Linn. Syft. Nat. 2. 444.—Fn. Suec.* 1405.—*Fabr. Ent. Syft. T. 3. p. 2. 324. 161.*

*Clerk Ic. t. 12. f. 12.*

This is a pretty, and by no means uncommon infect in England. Linnæus named it fpecifically Rhediella, in compliment to Rhedi, the author of feveral well known tracts on Natural Hiftory, that appeared about the end of the feventeenth century. The infect is figured in *Clerk's Icon.* a work executed under the immediate direction of Linnæus.

VOL. XI.	H	FIG.

# FIG. II.

## PHALÆNA ALBIDANA.

BROWN-DOTTED PALE TORTRIX-MOTH.

\* TORTRIX.

*SPECIFIC CHARACTER.*

TORTRIX ALBIDANA. Whitiſh: anterior wings with a double oblique tranſverſe ſeries of brown dots towards the poſterior end; an obſcure coſtal ſpot near the middle.

————————

This delicate little Inſect was taken in Coombe Wood, Surrey, and at Godſton in the ſame county. It is a tortrix of intereſting figure, though pale in colour. The wings are whitiſh: anterior pair faintly tinged with brown, and in addition to the double ſeries of brown dots towards the poſterior end of the wings, as mentioned in the ſpecific character, there are ſome other minute dots of the ſame colour ſparingly ſprinkled over the reſt of the anterior wings, and in particular two more diſtinct than the others appear in the diſk, a little inclining towards the baſe of the wing.

Not having obſerved the deſcription of this inſect in the work of any author, we ſhall venture to admit it as a new ſpecies.

FIG.

PLATE CCCLXXVII.      51

## FIG. III.

## PHALÆNA PROFANANA.

BROWN-TUFTED TORTRIX-MOTH.

*SPECIFIC CHARACTER.*

Wings fufcous grey, with a dark tufted dot in the middle.

Tortrix Profanana: alis cinereis: puncto medio fufco. *Fabr.*
*Ent. Syft. T. 3. p. 2. 268. fp. 111.*

———————

A new fpecies, defcribed by Fabricius as a native of this country, from a fpecimen in the cabinet of Mr. Francillon. Our infect, which was taken in Kent, is of the fame fpecies precifely, but only of a darker colour. Befides the confpicuous elevated hairy tuft in the middle of the wing, there are feveral minute elevated dots in the difk contiguous to it, as Fabricius mentions. The anterior wings have a rich filky glofs, the lower ones are brownifh, and imma-culate.

This, we believe, to be one of the rareft Britifh fpecies of that particular family of *Tortrices* known among Englifh collectors by the denomination of *Button-wing moths*, a term alluding to the fmall fafciculi, tufts, or fcabrous elevations, which appear on the anterior wings of fome few fpecies of the Tortrix tribe of Phalæna. It is alfo an infect of no very inconfiderable magnitude, as will ap-pear from the fmaller figure fhewn at number 3 in the oppofite plate, which reprefents it in the natural fize.

H 2          PLATE

# PLATE CCCLXXVIII.

## SCARABÆUS RURICOLA.

RUFOUS DARK-BORDERED SCARABÆUS.

COLEOPTERA.

*GENERIC CHARACTER.*

Antennæ clavated, the club fiffile : fhanks of the anterior legs generally dentated.

\* *Section* Melolontha, *mandible arched, and fomewhat dentated: wing-cafes fhorter than the body: naked extremity of the abdomen obliquely truncated.*

*SPECIFIC CHARACTER*

AND

*SYNONYMS*

Deep black, filky: wing-cafes rufous, marginal border, and future black.

SCARABÆUS RURICOLA: ater fericeus, elytris rufis: margine futuráque nigris. *Marfh. Ent. Brit. T. I. p.* 39. *fp.* 6.

MELOLONTHA RURICOLA: ater fericeus elytris rufis : margine nigro. *Fabr. Sp. Inf. I. p.* 73. *n.* 45.—*Mant. Inf. I. p.* 23. *n.* 58.—*Ent. Syft. I.* 173. *Sp.* 75.

SCARABÆUS RURICOLA: ater fericeus, elytris rufis; margine nigro. *Gmel. Linn. Syft. Nat.* 1558. *fp.* 235.

Scarabæus

Scarabæus niger, elytris croceis margine nigro.—Le Scarabé à bor-
dure.    *Geoffr. T. I. p.* 80. *fp.* 15.

Scarabæus marginatus.    *Fourc. I.* 9. 15.

Melolontha Floricola.    *Laich. I.* 41. 6.

---

Scarabæus Ruricola does not appear in either of the Entomological works of Linnæus. Fourcroy defcribes it as a Parifian infect \* ;
Fabricius as a native of England †; Roffius as an Italian fpecies ‡ ;
and we have a fpecimen of it from Germany; of which laft country,
Panzer gives it as an inhabitant in his *Entomologia Germanica.* We
are thus explicit, in order to fhew that it is a general European infect, and not exclufively a native of this country, as might be inferred
from the concluding obfervation of the Fabrician defcription of
this fpecies. " Habitat in Angliæ graminofis Dom Lee."

We muft acknowledge that, in the courfe of our own collecting, we
have never taken this infect, or feen it alive. Our figures are copied from an Englifh fpecimen, in the cabinet of that indefatigable
collector, the late Mr. Green, of Weftminfter, whofe cabinet has
recently fallen into our poffeffion, and where he met with it we cannot
afcertain. Mr. Marfham informs us *(Ent. Brit.)* that this infect
was taken in great abundance in the month of July, 1797, on New-
market Heath, near the Fofs, vulgarly called the Devil's Dyke.

There are two, if not more varieties of this infect, one of which
has the difk of the wing-cafes teftaceous inftead of rufous; Geof-

---

froy

froy even fays yellow " fes ètuis font *jaunes*, bordés de noir."
Fabricius fpeaks of another kind, in which the difk of the wing-
cafes is obfcure, with the furrounding border still darker. All
the under parts of this infect is black. Fig. I. fhews the natural fize.

**PLATE**

379

# PLATE CCCLXXIX.

## CHALCIS CLAVIPES.

THICK-LEGGED CHALCIS.

HYMENOPTERA.

*GENERIC CHARACTER.*

Mouth with a horny, compreſſed, and ſometimes elongated jaw: feelers four, equal: antennæ cylindrical, fuſiform, firſt joint rather thickeſt; thorax gibbous, lengthened behind, and obtuſe: abdomen ſmall, rounded, and ſubpetiolate: poſterior thighs thickiſh.

*SPECIFIC CHARACTER*

AND

*SYNONYMS.*

Black: thighs of the hind legs thick, and rufous.

CHALCIS CLAVIPES: atra, femoribus poſterioribus incraſſatis rufis. *Fabr. Mant. Inſ.* 1. *p.* 272. *n.* 2.—*Ent. Syſt. T.* 2. 195. *n.* 2.—*Hybn. Naturf.* 24. 56. 19. *tab.* 2. *fig.* 23.—*Roſs. Faun. Etruſc.* 2. 58. 803.—*Gmel. Syſt. Nat. T.* 1. *p.* 5. 2742. *n.* 2.

This very curious ſpecies of Chalcis is certainly undeſcribed as a Britiſh Inſect. The ſpecimen from which the figures in our plate are copied, and which is in our own cabinet, was taken in the vicinity of Faverſham, in Kent. This is not the only inſtance within our knowledge of its being caught in England; we find one ſpeci-

VOL. XI.                    I                    men

men of it in the Englifh cabinet of the late Mr. Drury. Independently of thefe, we have feen alfo two examples of it in the collection of T. Marfham, Efq. that were taken by himfelf in Kenfington Garden, fome years ago. Thefe are, however, the only Britifh fpecimens of Chalcis Clavipes we are acquainted with, from whence we may prefume to think it very far from common. Fabricius, upon the authority of Hybner, defcribes it as an inhabitant of Saxony; a fpecimen of it from France, occurs in the cabinet of A. M'Leay, Efq.

The fmalleft Figure denotes the natural fize.

**PLATE**

# 380

# PLATE CCCLXXX.

## FIG. I. I.

## PHALÆNA SUBOCELLANA.

SUB-OCELLATED TORTRIX-MOTH.

LEPIDOPTERA.

*GENERIC CHARACTER.*

Antennæ gradually tapering from the bafe to the tip : wings in general deflected when at reft. Fly by night.

\* TORTRIX.

*SPECIFIC CHARACTER.*

TORTRIX SUBOCELLANA. Anterior wings white, with fhort
oblique black lines at the exterior margin : bafe, and dufky fpot near
the tip, dotted with black ; a terminal gilt orange ftreak next the pofterior margin.

———————

Specimens of this infect have occured to our obfervation, in which
the black dots at the bafe of the anterior wings are fo intimately connected as to appear like interrupted tranfverfe lineations. The ground
colour is white : fometimes yellowifh ; and moft exquifitely mottled,
and dotted with black and dufky fpots, leaving the center of the difk
immaculate. The pofterior wings are pale.

This was taken in Kent, in the month of July. The fmalleft
figure denotes the natural fize.

I 2                                                    FIG.

## FIG. II. II.

## PHALÆNA MINISTRANA.

TESTACEOUS TORTRIX-MOTH.

*SPECIFIC CHARACTER*

AND

*SYNONYMS.*

Anterior wings teftaceous, with pofterior rufous margin; in the middle a ferruginous daub, and fmall white line.

TORTRIX MINISTRANA : alis anticis teftaceis : margine poftico rufo, medio litura ferruginea : lineola alba. *Linn. Syft. Nat.* 2. 877. 300.—*Fn. Suec.* 1131. *Gmel. Linn. Syft. Nat.* 2505. *n.* 300.

PHALÆNA MINISTRANA. *Fabr. Sp. Inf.* 2. *p.* 279. *n.* 20. *Mant. Inf.* 2. *p.* 227. *n.* 31. *Ent. Syft. T.* 3. *p.* 2. 252. *n.* 42.

———

Phalæna Miniftrana is rather an abundant infect in this country, for the moft part frequenting gardens. It is mentioned as a very common fpecies in Germany and Sweden, and moft probably is fo likewife throughout the reft of Europe.

**PLATE**

381

# PLATE CCCLXXXI.

## NAUCORIS CIMICOIDES.

### CIMEX-FORMED NAUCORIS.

#### GENERIC CHARACTER.

Snout fomewhat infleincluded: antennæ very fhort: lip advanced and rounded: wings four, folding crofs-wife: anterior legs cheliform.

#### SPECIFIC CHARACTER

Abdomen ferrated at the margin: head, and thorax, varied with yellow and brown.

NAUCORIS CIMICOIDES: abdominis margine ferrato capite thoraceque flavo fufcoque variis. *Geoffr. Inf. I.* 474. 1. *tab.* 9. *fig.* 5.—*Fabr. Spec. Inf. T.* 2. *p.* 334. *n.* 1. *Mant. Inf.* 2. 277. *Ent. Syft. T.* 4. *p.* 66. 210. *n.* 1.

NEPA CIMICOIDES, *Linn. Faun. Suec.* 907.—*Linn. Syft. Nat.* 2. 714. 6.—*Gmel. Linn. Syft. Nat.* 2122. *n.* 6.—*De Geer Inf.* 3. *p.* 375. *n.* 3. *t.* 19. *f.* 8. 9.

Cimex aquaticus latior. *Frifch. Inf.* 6. *p.* 31. *t.* 14.

La Naucore *Geoffr. Inf. I. p.* 474.

Naucoris

Naucoris Cimicoides is an inhabitant of the water, where it fubfifis by preying on a variety of other infects, which it attacks and pierces with its formidable, acutely pointed probofcis, and extracts their moifture in the fame manner as the bug, or cimex tribe. The habits of this animal are fimilar to thofe of the nepæ or water Scorpions, among which Linnæus places it, though not in our mind with fufficient reafon : we are perfuaded they ought, on the contrary, to conftitute two diftinct genera. Geoffroy was of this opinion : he feparated our infect from the nepæ, and referred it to his new genus *Naucore,* or Naucoris\*, and Fabricius follows the example of Geoffroy in this particular in his *Entomologia Syftematica.* It is a ftrong, and pretty evident characteriftic of the two genera Nepa, and Naucoris, that the firft has not the leaft appearance of a lip to the mouth, and the other has one very vifible and diftinct :—an advanced lip of a rounded form †.

This infect is not common. Our fpecimens were taken in Kent. It is well known as an European infect, though not as a Britifh fpecies.

---

\* *Hift. Abreg. des Infectes, &c.*

† The fpecies of the Naucoris genus from this circumftance might be extremely well diftinguifhed by the trivial Englifh name of *Round-Lipped Water-Scorpions* among the entomological collectors in this country ; the Nepæ are fimply *Water Scorpions.*

**PLATE**

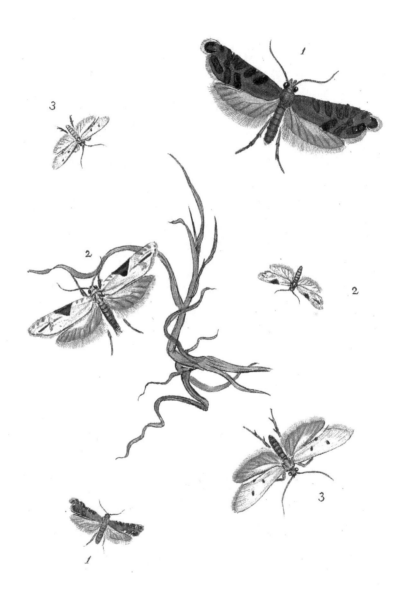

# 382

# PLATE CCCLXXXII.

## FIG. I. I.

## PHALÆNA TRIGUTTELLA.

TRIPLE SILVER-DOT TINEA-MOTH.

LEPIDOPTERA.

*GENERIC CHARACTER.*

Antennæ tapering from the bafe, wings in general deflected when at reft. Fly by night.

\* TINEA.

*SPECIFIC CHARACTER.*

PHALÆNA TRIGUTTELLA. Anterior wings brown, with oblique marginal filvery ftreaks: at the pofterior tip a black line, enclofing three filvery dots.

This we have taken, not very uncommonly, in woods, about the month of July.

## FIG. II. II.

## PHALÆNA HUBNERELLA.

HUBNER'S TINEA MOTH.

*SPECIFIC CHARACTER.*

TINEA HUBNERELLA. Anterior wings fublanceolate, pale, with a coftal triangular fufcous fpot.

Taken

Taken in the vicinity of Faverſham, Kent.   Not defcribed, to our knowledge, by any author.

---

## FIG. III. III.

### PHALÆNA TRIPUNCTELLA.

THREE-SPOT TINEA MOTH.

*SPECIFIC CHARACTER*

*AND*

*SYNONYMS.*

Wings cinereous, with three fufcous dots in the difk.

Tinea Tripunctella : alis cinereis; punctis tribus fufcis. *Fabr. Ent. Syſt. T. 3. p. 2. p. 312. ſp. 114.*

Tinea Tripunctella.  *Wien Verz. App.*

---

Defcribed by Fabricius from a ſpecimen in the cabinet of Schieffermyller as an inhabitant of Auſtria.  Our ſpecimen is from Faverſham, Kent.

PLATE

# 383

# PLATE CCCLXXXIII.

## PYROCHROA RUBRA.

### BLACK-HEADED CARDINAL-BEETLE.

#### COLEOPTERA.

*GENERIC CHARACTER.*

Antennæ filiform, with pectinated teeth : head exserted : thorax flat, orbicular, and immarginate : wing-cafes flexile : body oblong, and thickeft behind.

*SPECIFIC CHARACTER*

AND

*SYNONYMS.*

Black : thorax, and wing-cafes fanguineous, without fpots.

PYROCHROA RUBRA. *De Geer.* 5. 20. 1. *t.* 1. *f.* 14.

LAMPYRIS COCCINEA. *Linn. Faun. Suec.* 705 ?

PYROCHROA COCCINEA : nigra thorace elytrifque fanguineis immaculatis. *Fabr. Ent. Syft. T. I. p.* 2. 104. 70.— *Gmel.* 1886. 18.—*Marfh. Ent. Brit. T. I. p.* 364. *n.* 2.

———

The Pyrochroa rubra of De Geer is an extremely rare infect in this country. In its general afpect it bears a pretty ftrong refemblance to

VOL. XI.        K        another

another fpecies of Pyrochroa already figured in this work\*, our *P. coc-cinea,* and *P. rubens* of Fabricius.    On a flight comparifon, the differ-ence is however obvious, Pyrochroa rubra being rather larger; the fan-guineous colour of the fuperior furface is alfo fomewhat brighter, and the head of a deep black as in the lower furface of the body, while the head of our P. Coccinea is of the fame red colour as the thorax, and wing-cafes.—Notwithftanding thofe differences, it fhould be obferved, that fome doubts ftill remain whether they are diftinct fpecies: the two fexes of the fame fpecies, or only mere varieties.    We think them diftinct, but Fabricius, upon whofe authority principally they have been feparated by moft late writers, is not perfectly fatisfied that they are fo. The moft diftinguifhing feature of the two infects confifts in one hav-ing the head red, and the other black.

This infect is found on rotten willows.

---

\* Pl. 56. fig. 1.

PLATE

# 384

# PLATE CCCLXXXIV.

## SPHINX ASILIFORMIS.

### CLEAR UNDER-WING HAWK-MOTH.

#### LEPIDOPTERA.

##### GENERIC CHARACTER.

Antennæ fomewhat prifm-formed, and thickeft in the middle: tongue moft commonly exferted: feelers two: wings deflected.

\* *Section* **Sefia**; *wings entire: tail bearded: palpi two, reflected: tongue exferted, and truncated: antennæ cylindrical.*

##### SPECIFIC CHARACTER

AND

##### SYNONYMS.

Anterior wings fufcous, pofterior ones tranfparent: abdomen bearded, black, with three yellow bands.

SESIA ASILIFORMIS: alis anticis fufcis: pofticis feneftratis, abdomine barbato atro: cingulis tribus flavis. *Fabr. Ent. Syft. T. 3. p. 1. 383. 16.*

SESIA ASILIFORMIS. *Wien Schmetterl app. 305.*

SPHINX SESIA: alis primoribus fufcis, pofterioribus feneftratis, abdomine atro: cingulis tribus flavis. *Gmel. Linn. Syft. Nat. 2389. fp. 102.*

SPHINX TABANIFORMIS. *Naturf. 7. 110. 4.*

SPHINX ASILIFORMIS. *Turt. Syft. Nat. 3. p. 181.*

SPHINX ASILIFORMIS. *Haw. Lep. Brit. 69. p. 19.*

K 2

An

An extremely rare fpecies in England. We have a fpecimen of it in very fine condition in the cabinet of the late Mr. Drury, that was taken near London, on the poplar. Fabricius fpeaks of it as an inhabitant of the South of Europe.

The fmalleft figure reprefents it in the natural fize.

PLATE

# 385

# PLATE CCCLXXXV.

## FIG. I. I.

### APIS LAPIDARIA.

RED-TAILED BEE.

HYMENOPTERA.

*GENERIC CHARACTER.*

Mouth horny: jaw and lip membranaceous at the tip: tongue inflected: feelers four, unequal, filiform: antennæ short, and filiform: those of the female somewhat clavated: sting of the females and neuters pungent, and concealed within the abdomen.

*SPECIFIC CHARACTER*

*AND*

*SYNONYMS.*

Body of the female black, hirsute, with red tail: that of the male above black, hirsute, with red tail; face before the antennæ, and thorax at the base and apex yellow.

APIS LAPIDARIA: hirsuta atra, ano fulvo. *Linn. Fn. Suec.* 1712.— *Gmel. Linn. Syst. Nat. p.* 2782. *sp.* 44.

APIS LAPIDARIA. *Fabr. Ent. Syst. T.* 2. *p.* 320. *n.* 25. *mas.*

APIS ARBUSTORUM. *Fabr. Ent. Syst. T.* 2. *p.* 320. *n.* 24. *fem.*

APIS LAPIDARIA corpore *femineo* atro, hirsuto, ano rubro: corpore masculo supra atro, hirsuto, ano rubro; fascie, thoracisque basi et apice, flavis. *Kirby Apium Angl. T.* 2. *p.* 363. *n.* 106.

L'abeille

L'abeille noire avec les derniers anneaux du ventre fauves. Et,
L'abeille noire à couronne du corcelet citron, et extrémité du ventre
fauve. *Geoffr. Hiſt. Inſ. p.* 417. *n.* 21 & 22.

———————

This bee, according to ſome recent obſervations of the Rev. Mr.
Kirby, is to be conſidered as the neuter of Apis lapidaria, the female of which was figured in plate 108 of this work, and a variety β in
plate 88 at fig. 2.

In ſize and appearance it bears the neareſt reſemblance imaginable
to the Fabrician Apis arbuſtorum, which laſt Mr. Kirby aſcertains to
be the male of Apis lapidaria, a faƈt that would not eaſily have been
ſuſpeƈted, had not opportunities been afforded of attending to its habits and manners in its native haunts. Reaumur ſpeaks of thoſe
bees, with one or two citron coloured bands on the body, being found
in the ſame neſts with lapidaria. Mr. Kirby has alſo ſeen it enter
the nidus of that ſpecies, but what, as he obſerves, appears to remove
all doubts of their being the ſame ſpecies, he ſaw the ſuppoſed male
inſeƈt in the colleƈtion of the late Peter Collinſon, with a memorandum affixed to it ſpecifying that he had ſeen it conneƈted with Apis
lapidaria.

———————

## F I G. II.

### APIS MUSCORUM.

YELLOW-BODIED MOSS BEE.

*SPECIFIC CHARACTER.*

Hirſute, fulvous, abdomen yellow.

APIS MUSCORUM: hirſute fulva, abdomine flavo. *Linn. Faun.*
*Suec.* 1714.—*Gmel. Linn. Syſt. Nat. p.* 2782. *n.* 46.

PLATE CCCLXXXV. 71

Apis Muscorum. *Fabr. Ent. Syſt. T. 2. p.* 321. *n.* 31.

Apis senilis. *Fabr. Ent. Syſt. T. 2.* 324. *n.* 44. Muſcorum *var.*

Apis Muscorum: hirſuto-flaveſcens; thorace fulvo. *Kirby Ap. Angl. T. 2.* 317. 74.

Apis Muſcorum is one of the more common ſpecies of wild bee found in Europe. It frequents fields and meadows, where it forms a neſt compoſed of moſs, in cavities or holes juſt below the ſurface of the earth.

## FIG. III.

## APIS BARBUTELLA.

BARBUT'S BEE.

*SPECIFIC CHARACTER.*

Black, hirſute: anterior part of the thorax, with the ſcutel fulvous: abdomen ſubglobular, tail white.

Apis Barbutella: atra, hirſuta, ano albo; vertice, thorace, an-ticè, ſcutelloque, fulvis; abdomine ſubgloboſo. *Kirby Ap. Angl. T. 2. p.* 343. *n.* 93.

There appear to be more than one or two diſtinct varieties of this kind of bee. The Fabrician Apis autumnalis, Apis ſaltuum, of Panzer, and Apis monacha, of Chriſtius, according to Mr. Kirby, are

are all intended for the variety β of his male Apis Barbutella, which he diftinguifhes as having the thoracic band, fcutel, and bafe of the abdomen hirfute, with greyifh hairs. The defcriptions certainly accord with it fo exactly, that we cannot hefitate in admitting the opinion of Mr. Kirby to be correct. APIS AUTUMNALIS hirta, thorace cinerafcente : fafcia nigra; abdomine atro bafi cinerafcente, ano albo. *Fabr\**.—APIS SALTUUM hirfuta atra, thorace albo fafcia nigra abdomine antice anoque albis. *Panz. Faun. Inf. Germ.*——Another variety has the bafe of the thorax and tip obfcure yellowifh, and the abdomen immaculate at the bafe. Apis Barbutella is not very uncommon in the fummer time among flowers : the variety called by Fabricius Autumnalis, is feen moft commonly late in the year, and on thiftles chiefly.

---

* Defcribed as a German infect nearly allied to Apis ruderata from the cabinet of Smidt. " Nimis affinis certe A ruderatæ at duplo minor. Caput atrum. Thorax hirtus, cinerafcens fafcia inter alas atra. Abdomen hirtum bafi cinerafcens, in medio atrum ano lato albo. Pedes nigri tarfis piceis." *Fabr. Ent. Syft. T. 2. p. 324. 43.*

PLATE

# 386

# PLATE CCCLXXXVI.

## FIG. I. II.

## PHALÆNA FLAVO-STRIGATA.

### ORANGE-BANDED CARPET.

#### LEPIDOPTERA.

##### GENERIC CHARACTER.

Antennæ gradually tapering from the bafe : wings in general de-flected when at reft. Fly by night.

#### * Geometra.

##### SPECIFIC CHARACTER.

PHALÆNA FLAVO-STRIGATA. Wings pale, with deep yellowifh clouded tranfverfe bands, and an obfcure central dot on the anterior ones.

———————————

The natural fize of this infect is fhewn at Fig. 1. It is an elegant infect, and rather uncommon.

## FIG. III.

## PHALÆNA FUSCO-UNDATA.

### TESTACEOUS DARK-WAVED CARPET.

##### SPECIFIC CHARACTER.

PHALÆNA FUSCO-UNDATA. Anterior wings fubteftaceous, with ir-regular fufcous waved bands, and a few fufcous dots.

VOL. XI.         L         Nearly

Nearly allied to the infects known among Englifh Aurelians by the name of the July high flyer, in its general appearance and markings, but different in colour, and is in particular deftitute of the fmall white fpot on the band at the pofterior apex of the firft pair of wings. This infect is from Faverfham.

## FIG. IV.

## PHALÆNA BOMBYCATA.

### CHEVRON MOTH.

*SPECIFIC CHARACTER.*

PHALÆNA BOMBYCATA. Anterior wings pale and fufcous, variegated with yellowifh : a broad tranfverfe band of teftaceous lines, with a central dark chevron-like mark in the middle.

Found in the month of May, principally on the broom.

PLATE

# 387

# PLATE CCCLXXXVII.

## CICADA BIFASCIATA.

BIFASCIATED FROG-HOPPER.

HEMIPTERA.

### GENERIC CHARACTER.

Snout inflected: antennæ fetaceous: four wings, membranaceous: legs in general formed for leaping.

\* *Section* Cerçopis. *Lip abbreviated, truncated, and emarginated. Fabr.*

### SPECIFIC CHARACTER

AND

### SYNONYMS.

Yellowifh: wing-cafes fufcous, with two whitifh bands.

CICADA BIFASCIATA: flavefcens elytris fufcis: fafciis duabus albidis. *Linn. Syft. Nat.* XII. 2. *p.* 706. *n.* 11.

Cicada fufca, fafciis alarum binis albis. *Linn. Fn. Suec.* 1. *n.* 633.— 898.

Cicada bifafciata. *Gmel. Linn. Syft. Nat. T.* 1. *p.* 4. 2101. 11.

Cercopis 2 fafciata. *Fabr. fp. Inf.* 2. *p.* 330. *n.* 13.—*Mant. Inf.* 2. *p.* 275. *n.* 20.—*Ent. Syft. T.* 4. 56. 40. *Panz. Fn. Germ.* 7. *tab.* 20.

Cicada trifafciata, *De Geer, Inf.* 3. *p.* 186. *n.* 6. *t.* 11. *f.* 25 ?

L 2

This

This is a beautiful little fpecies of the Linnæan Cicadæ, and by no means common. It inhabits Sweden according to Linnæus: from Panzer, it appears to be a German infect, and it is alfo found in France. Our fpecimen was taken near Faverfham, in Kent.

The fmalleft figures in the annexed plate, denote the natural fize of this infect

PLATE

# 388

# PLATE CCCLXXXVIII.

## PHALÆNA MENDICA.

SPOTTED MUSLIN MOTH.

LEPIDOPTERA.

*GENERIC CHARACTER.*

Antennæ gradually tapering from the bafe to the tip: tongue fpiral: wings in general deflected when at reft. Fly by night.

\* *Bombyx.*

*SPECIFIC CHARACTER*

AND

*SYNONYMS.*

Wings of the male brown and obfcure: thofe of the female, white and pellucid, both dotted with black.

PHALÆNA MENDICA: alis deflexis nigro punctatis, femoribus anticis luteis. *Linn. Syft. Nat.* 2. 822. 47.—*Gmel. Linn. Syft. Nat.* 2423. *n.* 47.

PHALÆNA MENDICA. *Fabr. Ent. Syft.* 3. 452. *n.* 139. Mas cinero fufcus, fæmina albida punctis aliquot nigris. Femora antica barba lutea. Abdomen concolor. *ibid.*

PHALÆNA MENDICA. *Marfh. in Linn. Tranf. T.* 1. *p.* 72.

The

The fpotted Muflin Moth is one of our rareft fpecies of Phalæna in this country. The larva feeds fecurely from the intrufion of the Entomologift in marfhes and watery places, fubfifting entirely on aquatic plants, and is therefore fcarcely even met with, except in the winged ftate, which it affumes in May. The diffimilarity between the two fexes of this fpecies is altogether fo very remarkable, that it is only from an intimate acquaintance with the manners of the two infects in a ftate of nature, or the concurrent teftimony of many obfervers, that we could be induced to believe them both of the fame fpecies.

Fabricius, and Gmelin after him, fays, the larva is greenifh, hairy, with whirls of black dots, and yellowifh head. The figures in Efper, *T*. 3. pl. 42, reprefent the larva of a cinereous colour, verticillated with black dots, and tufts of ferruginous hairs. Other writers fpeak of the head and tail being red. Thofe different defcriptions may be eafily, however, reconciled by prefuming thofe authors had each noticed the larvæ at different periods of growth, or perhaps this diffimilarity may ferve to point out the difference between the two fexes, even in the larva ftate.

PLATE

# 389

# PLATE CCCLXXXIX.

## FIG. I. I.

## CURCULIO RUBER.

RED CURCULIO.

COLEOPTERA.

*GENERIC CHARACTER.*

Antennæ clavated, and feated on the fnout, which is horny and prominent: pofterior part of the head thick.

*SPECIFIC CHARACTER*

AND

*SYNONYMS.*

Reddifh-teftaceous: thorax grey: wing-cafes clouded with whitifh.

CURCULIO RUBER: rufo teftaceus, thorace grifeo, elytris nebulis albicantibus. *Marfh. Ent. Brit. T.* 1. *p.* 251. *fp.* 39.

This fpecies, though fmall, is interefting for its rarity. The general colour of the body is reddifh brown: head, and thorax fufcous: wing-cafes ftriated, fomewhat villofe, and banded with whitifh. Taken in Kent.

FIG

## FIG. II. II.

## CURCULIO LINEATUS.

LINEATED CURCULIO.

*SPECIFIC CHARACTER*

AND

*SYNONYMS.*

Fulvous : with three paler lines on the thorax.

CURCULIO LINEATUS : fuscus, thorace ſtriis tribus pallidioribus.
    *Linn. Syſt. Nat.* 616. 80.—*Faun. Suec.* 630.—
    *Gmel. Linn. Syſt. Nat.* 1784. 80.

CURCULIO LINEATUS.  *Fabr. Syſt. Ent.* 148. 111.—*Sp. Inf.* 1.
    189. 155.—*Mant. I.* 116. 206.—*Ent. Syſt. I. p.*
    2. 466. 302.

Curculio lineatus.  *Marſh. Ent. Brit. T. I. p.* 309. *ſp.* 206.

Curculio roſtro thoracis longitudine, thorace tribus ſtriis pallidioribus.
    —et Le Charanſon à corcelet rayé.  *Geoffr. I.* 283.
    13.
    *De Geer. Inf.* 5. *p.* 247. *n.* 35.
    *Schoeff. Icon. t.* 103. *f.* 8.

Lives chiefly in rotten willows, and plants of the diadelphous kinds.
This inſect we have occaſionally found pretty common.

FIG.

## F I G. III. III.

## CURCULIO RUFUS.

RUFOUS CURCULIO.

*SPECIFIC CHARACTER.*

Rufous : eyes, breaft, and abdomen anteriorly black.

CURCULIO RUFUS : rufus, oculis, pectore, abdomineque antice nigris. *Marfh. Ent. Brit. T. I. p.* 261. *fp.* 69.

CURCULIO VIMINALIS. *Fabr. Syft. Ent.* 145. 92.—*Sp. Inf.* 1. 184. 126.—*Mant.* 1. 110. 115.—*Ent. Syft.* 1. *p.* 2. 447. 243.

CURCULIO faltator Ulmi. *De Geer, Inf. V.* 260. 48. *t.* 8. *f.* 5.

Curculio rufus, femoribus pofticis craffioribus, elytris rufis. *Le Charanfon* fauteur brun. *Geoffr. Inf.* 1. *p.* 286. *n.* 19.

—————

A very dark coloured variety of this fpecies has been given already in this work, Fig. I. Pl. 249. The prefent figure is introduced in order to convey a more accurate idea of the general afpect of the infect. They are occafionally found to vary from a pale yellowifh, or clay colour, to deep rufous, but the moft frequent variety is that now reprefented. Found on the nut tree.

# 390

2

1

# PLATE CCCXC.

## FIG. I.

## SCARABÆUS AGRICOLA.

### AGRICOLA BEETLE.

#### COLEOPTERA.

##### GENERIC CHARACTER.

Antennæ clavated, the club fiffile : fhanks of the anterior legs generally dentated.

\* *Section* Melolontha, *mandible arched, and fomewhat dentated: wing-cafes fhorter than the body : naked extremity of the abdomen obliquely truncated.*

##### SPECIFIC CHARACTER

###### AND

###### SYNONYMS.

Braffy black : thorax villofe : wing-cafes livid, with a black border, and arched band.

SCARABÆUS AGRICOLA : nigro-æneus, thorace villofo, elytris lividis : limbo fafciâque arcuatâ nigris, *Marfh. Ent. Brit.* T. I. p. 48. fp. 76.

SCARABÆUS AGRICOLA. *Linn. Syft. Nat. 2. 533. 58.*

M 2

MELO-

MELOLONTHA AGRICOLA: thorace villofo, elytris lividis: limbo fafciaque nigris, clypeo apice reflexo. *Fabr. Syft. Ent.* 37. 29.—*Sp. Inf. I.* 43. 44.—*Mant. I.* 23. 57.—*Ent. Syft. I.* 173. 74.

SCARABÆUS AGRICOLA. *Donov. Tour of South Wales. A. D. 1801—1804. Vol. 2. p.* 239.

Le Cyathiger. *Scop.* 6.

———————

About the latter end of the month of July, 1801, we were fo for-tunate as to capture a living fpecimen of Scarabæus Agricola, on the fea coaft of the county of Caermarthen, South Wales, thereby afcertaining, beyond a doubt, the exiftence of this lovely infect in our own ifland. As an European fpecies, it was well known be-fore the time of Linnæus, and has been fince mentioned by vari-ous continental authors, but no writer has hitherto fpoken of it as a native of this country, with the exception of Mr. Marfham, who, on our authority, inferted this fpecies in his recent publication, *En-tomologia Britannica.* The difcovery of this infect we may con-fider, therefore, of fome moment to the Entomologift, if not a va-luable acceffion to the Britifh *Fauna.*

The beauty of this infect, when alive, was eminently ftriking: the thorax did not appear of that obfcure dufky hue obfervable in dead fpecimens: both that part of the thorax, and the head, were of a rich braffy-green colour, and flightly villous; and the dark margin, with the band acrofs the wing-cafes, though black, were elegantly gloffed with purple.

FIG

PLATE CCCXC. 85

# FIG. II.

## SCARABÆUS FRISCHII.

FRISCH'S BEETLE.

\* *Section* Melolontha.

SPECIFIC CHARACTER,

AND

SYNONYMS.

Braffy black, gloffy : wing-cafes teftaceous.

MELOLONTHA FRISCHII: nigro ænea nitida elytris teftaceis. *Fabr.*
*Syft. Ent.* 37. 25.—*Sp. Inf. I.* 41. 35.—*Mant. I.*
21. 40.—*Ent. Syft. I. p.* 2. 167. 53.

SCARABÆUS FRISCHII: æneus, elytris teftaceis : futurâ virefcenti.
*Marfh. Ent. Brit. T. I. p.* 40. *fp.* 71.

SCARABÆUS FRISCHII: *Donov. Tour of South Wales, Vol. I.*
*p.* 377.

---

The only fpecimen of Scarabæus Frifchii that we have yet feen alive,
was obferved crawling upon a meagre blade of the common mat-weed
*Arundo arenaria,* that had fecured itfelf a local habitation upon a
fand-hill on the fea-coaft, about a mile to the weftward of Newton
Bay, Glamorganfhire. The whole body, except the wing-cafes (which
are teftaceous) the lateral edges of the thorax, legs, eyes, and antennæ,
were of a fine braffy purple. In fome fpecimens, the head, thorax,
fcutel, and tail, are of a coppery green inftead of purple.

Mifs

Mifs Hill found Scarabæus Frifchii among marine *rejeclumenta,* near Braunton Burroughs, Devonfhire, on the fhore of the Severn fea, nearly oppofite the county in which we difcovered it, and in a fomewhat fimilar fituation.    *Vide Marfh. Ent. Brit.*

PLATE

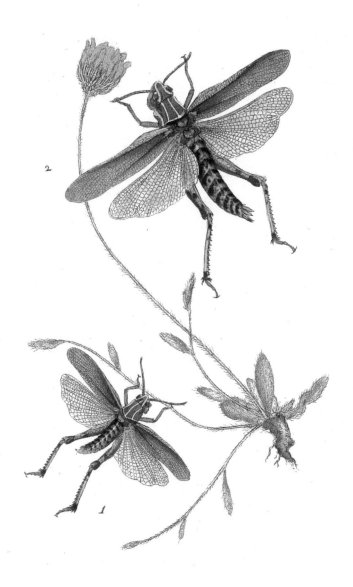

# 391

# PLATE CCCXCI.

## GRYLLUS FLAVIPES.

YELLOW-LEGGED GRASS-HOPPER

HEMIPTERA.

### GENERIC CHARACTER.

Head inflected, armed with jaws: feelers filiform: antennæ ufually fetaceous, or filiform: wings four, deflected, convolute, the lower ones plaited: pofterior legs formed for leaping: claws double.

* *Section* Gryllus. *Antennæ filiform: feelers equal, and filiform: tail fimple.*

### SPECIFIC CHARACTER

AND

### SYNONYMS.

Thorax above, and wing-cafes brown: anterior margin greenifh-yellow: pofterior thighs yellow, beneath fanguineous; fhanks yellowifh.

GRYLLUS FLAVIPES: thorace fupra elytrifque brunneis: margine anteriori viridi-flavis, femoribus pofterioribus fubtus fanguineis tibiifque flavis. *Linn. Syft. Nat. Gmel. Linn. Syft. Nat.* 2088. *n.* 230.

———

We are much inclined to fufpect that this beautiful fpecies of Gryllus has not been delineated by any author, although it is one of the
Linnæan

Linnæan infects, and on that account more likely to have been noticed than many others. Linnæus faw it in the mufeum of Lefk, from whence he defcribes it very accurately as an European infect, in his Syftema Naturæ, and it appears likewife in *Mufeum Lefkeianum.* It is rather fingular that Fabricius does not mention it.

This is probably a rare fpecies on the continent, or at leaft it would appear fo from the filence of continental Entomologifts refpecting it. As a Britifh Infect *Gryllus flavipes* is uncommon, having been hitherto found only by one or two collectors : they met with it in the vicinity of London, and fpeak of it as a fpecies peculiar to marfhy places. Found about the latter part of July.

Both fexes are reprefented in our plate in their natural fize, and in a flying pofition in order to difplay their wings in the moft picturefque point of view.

PLATE

# 392

# PLATE CCCXCII.

## FIG. I. I.

### PHALÆNA CRAMERELLA.

CRAMER'S TINEA-MOTH.

LEPIDOPTERA.

*GENERIC CHARACTER.*

Antennæ tapering gradually from the bafe: wings in general deflected when at reft. Fly by night.

\* TINEA.

*SPECIFIC CHARACTER*

AND

*SYNONYMS.*

Wings filvery, with three tranfverfe, brownifh, golden bands; and a black fubocellated dot at the tip.

TINEA CRAMERELLA: alis argenteis: lineolis obliquis marginalibus fafciis punctoque apicis atro. *Fabr. Ent. Syft. T. 3. p. 2. p. 327. fp. 173.*

This we have every reafon to believe muft be the infect intended by Fabricius for his fpecies Cramerella, allowing that his defcription was taken from a wafted fpecimen, in which the tranfverfe bands acrofs the wings appeared lefs diftinct than in the infect we have figured. He fpeaks

VOL. XI.  N  of

of it as a minute fpecies, and as an inhabitant of England. We have frequently taken it on ferns, and low herbage at the fkirts of woods.

---

## FIG. II. II.

### PHALÆNA BLANCARDELLA.

BLANCARD'S TINEA MOTH.

*SPECIFIC CHARACTER*

AND

*SYNONYMS.*

Wings golden, with a filvery fpace at the tip, and feven marginal fpots.

TINEA BLANCARDELLA: alis auratis: lineola apicis maculifque feptem marginalibus argenteis. *Fabr. Ent. Syft.* T. 3. p. 2. p. 327. *fp.* 175.

---

Defcribed by Fabricius as an Englifh infect, from the cabinet of Yeats. This has the fame haunts as the preceding, and is equally common.

---

## FIG. III. III.

### PHALÆNA EMARGINELLA.

CINEREOUS NOTCH-WING TINEA.

*SPECIFIC CHARACTER.*

TINEA EMARGINELLA. Anterior wings linear, and deeply emarginated at the outer edge: greyifh: apex ftreaked with fufcous.

This

This fpecies is found in Kent; it is very rare, and apparently undefcribed by any author.  There are feveral fpecies of the Tortrix family that have the outer edge of the anterior wings emarginate, and are known among Englifh collectors by the trivial epithet of Notch-wing Moths, but we do not recollect to have obferved the fame circumftance in any of the Tinea tribe before : fome few of the fmaller fpecies have the edges of the wings jagged, or indented, but not exactly in this manner.

N 2

PLATE

# PLATE CCCXCIII.

## CERAMBYX SCALARIS.

### YELLOW INDENTED-LINE CERAMBYX.

#### COLEOPETRA.

*GENERIC CHARACTER.*

Antennæ fetaceous: eyes lunate, and embracing the bafe of the antennæ: thorax generally fpinous, or gibbous: wing cafes fomewhat linear.

&ast; *Section* Saperda. *Thorax unarmed, fubcylindrical.*

##### SPECIFIC CHARACTER

###### AND

*SYNONYMS.*

Black, with an indented futural yellow line, and yellow dots on the wing cafes: antennæ moderate.

CERAMBYX SCALARIS: mutico fubcylindrico, coleoptris linea futurali dentata, punctifque flavis, antennis mediocribus. *Linn. Syft. Nat.* 632. 55.—*Gmel. Linn. Syft. Nat. T.* 1. *p.* 4. 1837. 55.—*Marfh. Ent. Brit. T.* 1. 329. *n.* 8.

CERAMBYX FLAVOVIRIDIS. *De Geer, Inf.* 5. *p.* 77. *n.* 14.

LEPTURA SCALARIS. *Linn. Faun. Suec.* 697.

SAPERDA SCALARIS. *Fabr. Syft. Ent.* 184. 2.—*Sp. Inf.* 1. 231. 2.—*Mant.* 1. 147. 2.—*Ent. Syft.* 307. 2. *Panz. Ent. Germ.* 256. 2.

This

This very beautiful infect is recorded as a Britifh fpecies of Ce-rambyx, upon the authority of T. Swainfon, Efq. of the Cuftom-houfe, who found a fpecimen of it in Dover-place, Surrey, fome years ago. Previous to that time it was perfectly well known to the conti-nental naturalifts as an European infect, but it does not appear to be confidered as a common fpecies in any country.

The fmalleft figure fhews the natural fize of this infect : an enlarged reprefentation is alfo given in order to exhibit the fpecies to more ad-vantage.

PLATE

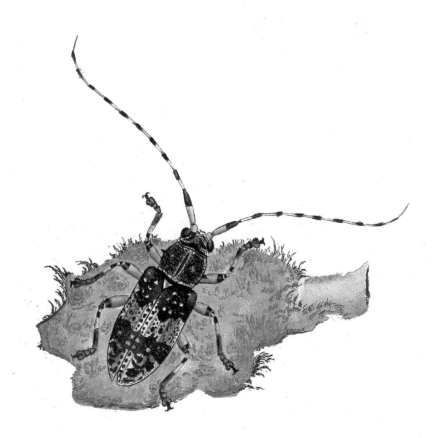

# 394

# PLATE CCCXCIV.

## CERAMBYX NEBULOSUS.

### CLOUDED CERAMBYX.

#### COLEOPETRA.

*GENERIC CHARACTER.*

Antennæ fetaceous: eyes lunate, and embracing the bafe of the antennæ: thorax generally fpinous, or gibbous: wing-cafes fomewhat linear.

*SPECIFIC CHARACTER*

*AND*

*SYNONYMS.*

Thorax fpinous: wing-cafes faftigiate, dotted, and banded with black: antennæ long.

CERAMBYX NEBULOSUS: thorace fpinofo, elytris faftigiatis: punctis fafciifque nigris, antennis longioribus. *Linn. Syft. Nat.* 627. 29.—*Fn. Suec.* 650.—*It. gotl.* 173. —*Gmel. Linn. Syft. Nat. T.* 1. *p.* 4. 1821. *fp.* 29.

CERAMBYX NEBULOSUS. *Fab. Syft. Ent.* 168. 20.—*Sp. Inf.* 1. 215. 26.—*Mant.* 1. 134. 36.—*Ent. Syft. T.* 1. *p.* 2. 261. 35.

CERAMBYX NEBULOSUS. *Marfh. Ent. Brit. T.* 1. *p.* 325. *fp.* 2.

Le Capricorne noir marbré de gris.—Et *Cerambyx* niger, elytris vellere cinereo marmoratis, antennis pedibufque cinereo interfeCtis. *Geoffr.* 1. 204. *fp.* 7.

Cerambyx

Cerambyx nebulofus is an infeⅽt of interefting figure, and under the lens of an opake microfcope, appears very elegant. Geoffroy in his *Hiſtoire Abrégeé des Inſeⅽtes*, obferves that it has been found upon willows. Fabricius and Gmelin fay it lives in the trunks of pines, and it is afferted to be highly injurious to the bark and timber of thofe trees in fome countries. It is not one of our rareft infeⅽts in England, though far from common.

PLATE

# 395

# PLATE CCCXCV.

## LEPTURA DORSALIS.

### YELLOW BROAD-BANDED LEPTURA.

#### COLEOPETRA.

*GENERIC CHARACTER.*

Antennæ fetaceous: head exferted: eyes roundifh, or oval and not embracing the bafe of the antennæ: thorax roundifh, attenuated in front, and fometimes fpinous or toothed: body oblong.

*SPECIFIC CHARACTER*

AND

*SYNONYMS.*

Black: wing-cafes with a broad tranfverfe yellow band: tip and two marginal fpots ferruginous.

LEPTURA DORSALIS: nigra, elytris flavo-fafciatis: apice maculifque duabus marginalibus ferrugineis. *Marfh. Ent. Brit. T.* 1. *p.* 343. 7.

―――――――――――

Leptura dorfalis was firft difcovered near Manchefter, by Mr. Philips, of that place. It is a rare infect, and feldom to be met with in Englifh cabinets. Mr. Marfham defcribes the male as being eight lines in length, and the female ten.

This is a very elegant fpecies, and has not been figured by any author.

## F I G. II.

## LEPTURA LAMED,

### FLEXUOUS-STRIPE LEPTURA,

#### SPECIFIC CHARACTER.

Thorax fpinous, pubefcent: wing-cafes faftigiate, livid, with a narrow ftripe down the middle, and a fpot behind, dufky.

STENOCORUS LAMED: thoraco fpinofo pubefcente, elytris faftigiatis lividis: tænia obfcura longitudinali finuata.
*Fabr. Ent. Syft. T.* 1. *p.* 2. 293. 82. *n.* 2.

STENOCORUS LAMED. *Panz. Faun. Inf. Germ.*

———————

This is one of our rareft, and moft interefting fpecies of Britifh Lepturæ. The fpecimen from which our figure is taken, we found in the Englifh cabinet of the late Mr. Drury. It has not been hitherto defcribed or noticed by any author as a Britifh infect.

PLATE

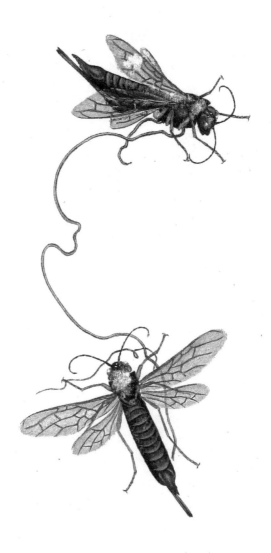

# 396

# PLATE CCCXCVI.

## SIREX JUVENCUS.

### BLUE-BODIED TAILED-WASP.

#### HYMENOPTERA.

*GENERIC CHARACTER.*

Mandible thick, horny, truncated at the tip, and denticulated: jaw incurvated, pointed, cylindrical, and ciliated: feelers four, pofterior pair longeft, and thicker towards the extremity: antennæ filiform, of more than twenty-four equal joints: fting exferted, ftiff, and ferrated: abdomen feffile, and terminating in a point: wings lanceolate.

*SPECIFIC CHARACTER.*

AND

*SYNONYMS.*

Abdomen deep blue: head and thorax greenifh black, and villous.

SIREX JUVENCUS: abdomine atro-cærulefcente, thorace villofo unicolore. *Linn. Faun. Suec.* 1575.—*Linn. Syft.
Nat.* 2. 929. 3.—*Gmel. Linn. Syft. Nat.* 2672.
*Fabr. Spec. inf.* 1. *p.* 419. *n.* 6.—*Mant. inf.* 1.
*p.* 257. *n.* 8.—*Ent. Syft. T.* 2. *p.* 126. 9.

Urocerus. *Schæff. icon. t.* 205. *f.* 3.
*Sulz. hift. Inf. t.* 26. *f.* 9. 10.

———

We poffefs an Englifh fpecimen of this very fcarce and beautiful infect, *Sirex Juvencus,* in the cabinet of the late Mr. Drury, but
whether

whether he caught it himfelf, or in what manner he obtained it, is
entirely unknown to us.—However, a fecond fpecimen of the fame
infect, upon which we may be allowed to fpeak with more confidence,
was lately communicated to us by Mr. Milton, engraver ; who caught
it as it refted againft a window in one of the upper apartments of his
houfe in Martlett's Court, Bow Street.—It is not undeferving of remark to the early entomologift, that many of our moft choice, and
rare infects, have occurred by accident in fimilar fituations, and have
never perhaps been found in any other.

This infect inhabits woods.  There is a variety of it with yellow
antennæ inftead of black, and which fometimes has the legs yellowifh
inftead of rufous.

# LINNÆAN INDEX

TO

## VOL. XI.

### COLEOPTERA.

# INDEX.

## HEMIPTERA.

## LEPIDOPTERA.

Phalæna

# I N D E X.

## HYMENOPTERA.

## DIPTERA.

P 2

# ALPHABETICAL INDEX

## TO

## VOL. XI.

# I N D E X.

Sphecoides,

# I N D E X.

## ERRATUM.

Page 23, l. 2. *For* Synodendron cylindricus *read*
Synodendrom cylindricum.

Printed by Bye and Law, St. John's Square, Clerkenwell.